Radio Tracer Techniques for Agriculturists & Biologists

NIPA® GENX ELECTRONIC RESOURCES & SOLUTIONS P. LTD.
New Delhi-110 034

www.nipaers.com
Browse, Search, Read & Buy...
Print Books
eChapters
eBooks
Publishers
Forthcoming
Subject
Catalogues
New Titles
Online Resources on:
✓ Current Affairs
✓ Reasoning, Logic & Aptitude
✓ Competitive Examinations
✓ English Language Lab
✓ Writing & Pronunciation Tools
✓ Personality Development
✓ Online Programmes for Professionals
✓ Interview Preparation
Pay Using
PhonePe
Google Pay
paytm
PayPal
UPI
CCAvenue

Radio Tracer Techniques *for* Agriculturist & Biologists

Authors:

B. A. GOLAKIYA, Professor & Head
H. L. SAKARVADIA, Agril. Officer
POONAM I. JETPARA, Junior Research Fellow
K. B. PARMAR, Agril. Officer
S. V. PATEL, Asso. Research Scientist

Department of Biochemistry, Agricultural Chemistry and Soil Science
Junagadh Agricultural University, Junagadh, Gujarat, India

NIPA® GENX ELECTRONIC RESOURCES & SOLUTIONS P. LTD.
New Delhi-110 034

NIPA® GENX ELECTRONIC
RESOURCES & SOLUTIONS P. LTD.

101,103, Vikas Surya Plaza, CU Block
L.S.C. Market, Pitam Pura, New Delhi-110 034
Ph : +91 11 27341616, 27341717, 27341718
E-mail: newindiapublishingagency@gmail.com
web: www.nipabooks.com

For customer assistance, please contact
Phone: + 91-11-27 34 17 17
Fax: + 91-11- 27 34 16 16
E-Mail: feedbacks@nipabooks.com

ISBN: 978-81-19254-52-1

Composed and Designed by NIPA®.

Dr. B. K. Kikani
Vice Chancellor

Junagadh Agricultural University
Junagadh

FOREWORD

Radio Tracer Techniques are advancing fastly for majority of disciplines. Biosciences are not exemption in it. Radio Tracer Techniques has professional and research and development uses. Knowledge of Radio Tracer Techniques employed in the bioscience will help plan the research projects accordingly. It can also help in the ongoing research projects. It is fast and accurate. Of course, the set of the Radio Tracer Techniques Lab. is a bit costly affair. But looking to the rapid and accurate completion of work, on the whole it becomes cost effective.

Several generations of the scientists have been scarified in the development of Radio Tracer Techniques especially for biosciences. Now it has become a full flagged branch. It is a time to be aware of the Radio Tracer Techniques especially for life sciences people.

Our group of scientist of the Department of Biochemistry has made praise worthy effort while recollecting and reorganizing the literature related to Radio Tracer Techniques. It has been compiled in a lucid and users friendly mode. I hope this collection will be cheered up by the all concerned with biosciences.

- Sd -

(B. K. Kikani)

Place: Junagadh

Dr. D. B. Kuchhadia
Director of Research and
Dean of P. G. Faculty

Junagadh Agricultural University
Junagdh

MESSAGE

Innovations of radio isotopes and its irradiations formed mile stones in the history of science. It consumed entire lives of several scientists. Of course this branch of science has been appreciated with half of dozen of Nobel Prices in the course of time. It is their rigorous efforts, labour and pain put behind understanding cause and effect relations of radiations emitted from radio isotopes; has resulted in to a full flagged branch of Radio Tracer Techniques. A lot of instrumental method development has been done in aspect. We are more concerned with the Radio Tracer Techniques in biosciences.

Biosciences are more intricate than other branch of sciences. Radio Tracer Techniques are employed in agriculture, entomology, zoology, marine science, pathology, ecology, ornithology, soil science, soil chemistry, animal husbandry, nutrient metabolism, medical science, food metabolism, disease diagnosis, drug metabolism and biotechnology.

Radio Tracer Techniques basically trace out the path. It may be a migration of birds, marine beings, reptile, and wild animals or may be the path of a compound or element within the living system or outside it. The duration and detection limits depend on the half life of an isotopes and the precision of the detecting equipments. But it is much more precise and accurate than the classical methods.

Certain detections and tracing are almost impossible with the conventional methods. Radio Tracer Techniques made them possible through novel principles and practices. Biosciences have been made a quantum jump after the innovations and implementation of Radio Tracer Techniques.

Our team at the Department of Biochemistry, JAU, Junagadh has made a remarkable effort while recollecting, recognizing and presenting the literature of Radio Tracer Techniques in Biosciences under a single volume. I am sure that the personal affiliated with bioscience will find it interesting and useful all the while.

- Sd -

Place: Junagadh

(D. B. Kuchhadia)

PREFACE

When I was doing my Ph.D. none of the teachers were ready to teach me course of Radio Tracer Techniques. The reason was it includes basis of molecular physics, application in various disciplines and laboratory equipments their principles and usage. This was considered as a tough subject. I preferred to study it only because it was a tough task. A dummy teacher was nominated and I started preparing this course. I was standing in the library even up to eight hours, canning over the racks of books in various disciplines in the library. Ultimately 45 standard books and several dozen back volumes of world class journals were selected and I brood over it continuously for four months and the first manuscript of this book was ready before 20 years. The compilation at the vary initial stage received wide coverage. My handouts were circulated in several universities. By chance I joined as a faculty in the same department where I was once a student. Automatically, I shouldered the responsibility of teaching Radio Tracer Technique. I am teaching this subject since last 20 years. Every year my students also suggested so many points and we were enriching the content of our manuscript. My students have taken keen interest in this particular subject. Rather they have been instrumental in shaping this script during two decades. It rarely happens that any book before coming into black and white takes such a long incubation period. While during the years long interaction this text has become more palatable, lucid and interesting. We hope this collection is a time tested work and will be appreciated by students and teachers as well.

Authors

List of Figures

List of Tables

CONTENTS

Chapter 1

Structure of the Atom

Even though we are now in the atomic age, the idea of the atom is not new. More than 2,000 years ago, the Greek philosopher Democritus suggested that any matter is composed of tiny indestructible particles called Atoms. The word atom is derived from the Greek word Atoms meaning something that can not be cut further. However not much attention was paid to this concept till 1808, when Dalton proposed the famous "Atomic theory". According to this theory, "all matter is composed of tiny indivisible particles called 'Atom'. Combining the ideas of Lavoisier, Boyle and his own, Dalton proposed that atoms of

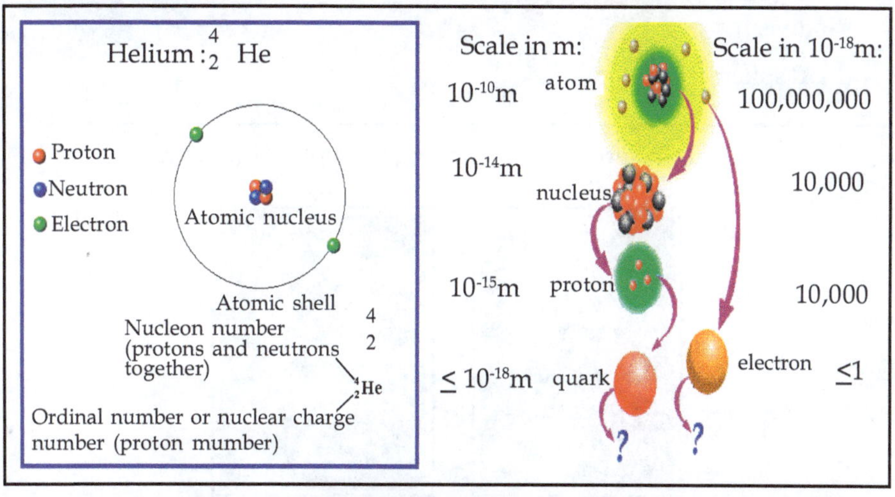

Fig. 1.1: Structure of helium atom

elements combine together to make molecules of a compound. This was a great step in the concept of the atom and compounds, since this gave a clear picture of how the supposedly smallest particles of matter called atoms could be put together to form the various familiar substances found in nature.

1.1 Changing Concepts of the Atom

Thomson's atomic model

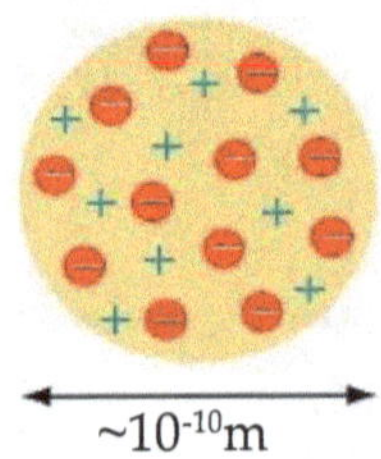

Thomson (1897) who was the first to propose the model of the atom (plum pudding model) postulated that an atom consisted of a homogeneous sphere, of positive charge (pudding) with the negative charged electrons (the plums) embedded uniformly in such a manner that their mutual repulsion was balanced by the mass attraction between them. The number of units of positive charge was presumed to be equal to the number of electrons so that the atoms were outwardly neutral. The spectrum of light emitted by the atom was attributed to vibrations of electrons about the mean position.

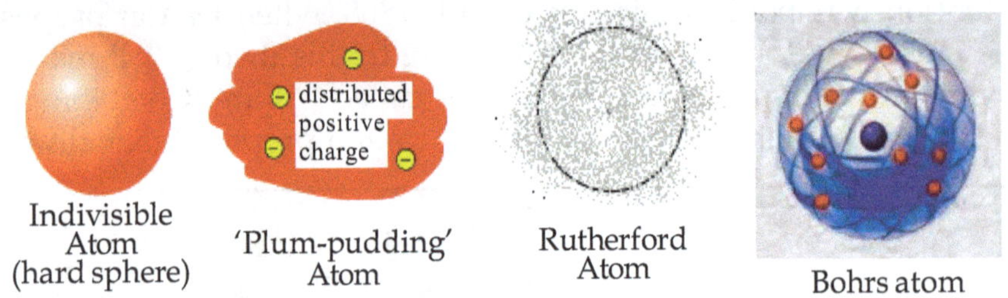

Fig.1.2: Periodically changing concepts of the atom

Thomson

Rutherford

Rutherford proposed the Dynamic Planetary Model wherein the positive charge of the atom must be concentrated in a very small region, less than 10^{-12} cm in diameter, and that almost the entire mass of the atom must be associated with the central core which he called the nucleus. The electrons were assumed to exist outside the nucleus. To explain the stability of the atom, he assumed that the electrons must revolve round the nucleus like planets around the sun so that the "centrifugal force may balance the electrostatic attraction between the nucleus and the electrons". However, this was contradictory to the classical electromagnetic theory, according to which an electron moving in a circular path would radiate energy and such emission would result in the loss of energy. Due to this continuous loss of energy, in the course of time it should approach the nucleus along a spiral path and ultimately fall into the nucleus. Hence there could not be any atomic stability.

However Rutherford showed theoretically that the number of particles that were scattered in any direction must be proportional to $[Ze]^2$ where Z is the positive charge in the nucleus and *e* is the unit charge. Thus the concept of atomic number of an element was introduced and was thought of as

1. The ordinal number of the element in the periodic table starting with hydrogen as 1,
2. The number of units of positive charge in the nucleus, and
3. The normal number of electrons outside the nucleus. Thus the atomic number Z was from then on always a whole number.

A black big ball at the center is the nucleus and small - points moving around the nucleus are electrons.

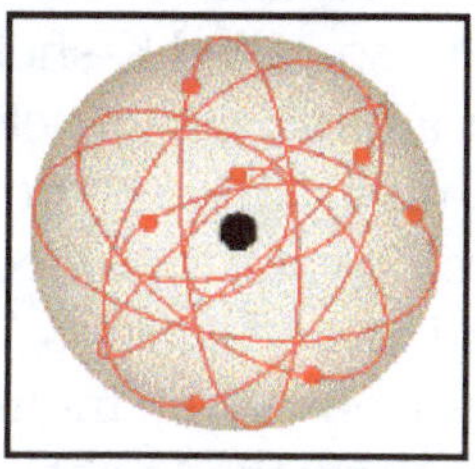

The Rutherford Model of the nuclear Atom

To overcome the objections raised to Rutherford's nuclear model of the hydrogen atom. Niel Bohr postulated that the electrons revolving in a circular orbit need not radiate energy and also need not have the possibility of revolving in all orbits but only in a selected number thereof. He called these allowed orbits stationary orbits in which the electrons could revolve for any length of time without emitting radiation. He further postulated that an electron could jump from one stationary orbit of higher energy to another of lower energy emitting a quantum of energy as radiation during the transition. However, even this model could not explain several other phenomena like the multiple structures of spectral lines. Arnold Summerfield introduced the idea of elliptic orbits instead of circular orbits and applied corrections for the variations of the electron mass with velocity. Subsequently Pauli proposed the exclusion principle (latter known as Pauli's principle) which stated that no two electrons in the same atom could have the same quantum numbers. Based on this and further studies, the following structure was proposed for the nucleus and the extra nuclear electrons in a neutral atom.

Niel Bohr

An atom was considered to consist of a positively charged nucleus containing as many protons as the nearest whole number (A) to the atomic weight and (A-Z) electrons, where Z is the atomic number. Thus, the nucleus has a surplus positive charge of Z units. Outside the nucleus, the Z numbers of electrons were arranged in a number of groups of shells so that the atom outwardly was electrically neutral. These extra nuclear electrons were arranged as per the Pauli's principle. Each shell with a total quantum number n could have a maximum of $2n^2$ electrons.

Based on the above structure of the atom, rational explanation of the periodic classification was given. In the case of the hydrogen atom, there is one proton in the nucleus and one electron in the inner most so-called K-shell. Helium which has an atomic weight close to 4 and atomic number 2 was thought to have four protons and two electrons in the nucleus and two electrons in the K-shell. The next shell L is occupied by electrons of elements starting with lithium and ending with neon. Similarly the M-shell is occupied by the electrons of elements beginning with sodium and ending with argon. The number of electrons in the outermost shell usually determines the valences of the element and its groups in the periodic table. One or more shells are always found to be completed with an inert gas and with the exception of helium; every inert gas has eight electrons in its outermost shell.

A mass number; Z atomic number; $^{14}_{6}C$

N neutron number = A - Z

1.2 Discovery of the Neutron

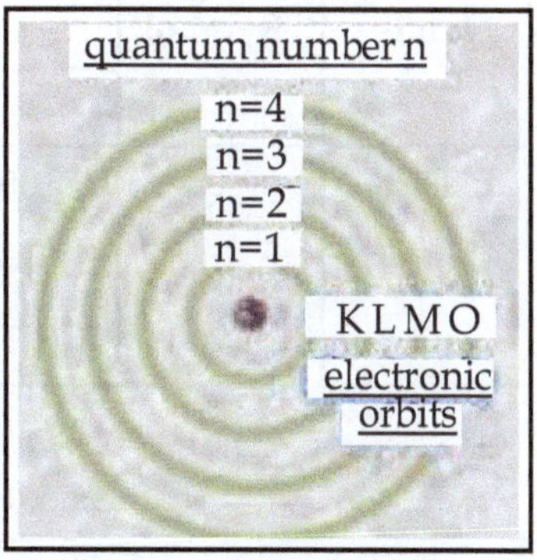

However, the concept of a proton-electron atomic nucleus was found to be inconsistent with the small size of the nucleus. Already in1920, Rutherford predicted the existence of a nuclear particle having mass but no charge. He said; "Under some conditions, it may be possible for an electron to combine more closely with a proton forming a kind of "neutral doublet". He called this hypothetical particle a neutron. Nevertheless the actual discovery of the neutron came much later and gave a new picture of the atomic nucleus.

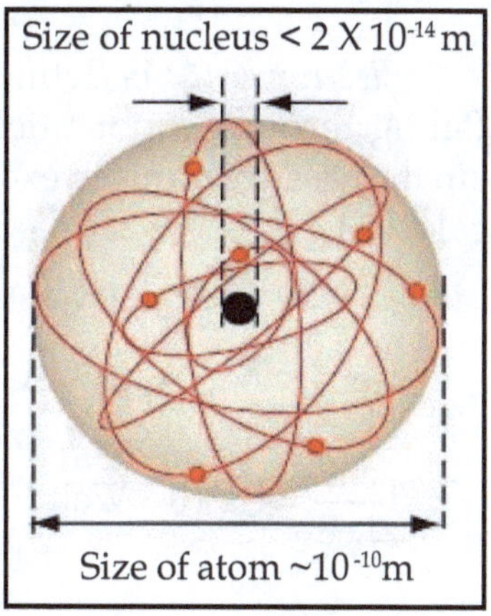

As such, the atom is now considered to be a positively charged nucleus containing singly charged; positive protons (p) each with a mass of 1.0076 a.m.u.(atomic mass units) and uncharged neutrons (n) each with a mass of 1.0089 a.m.u. held together in the nucleus by a force called the binding energy. The positively charged nucleus is surrounded in a circular or elliptical orbit by one or more electrons. For example, a gram of hydrogen contains 6×10^{23} atoms and the mass of an atom of hydrogen is 1.6 a.m.u. The diameter of the atom is about 10^{-8} cm whereas that of the nucleus is about 10^{-12}cm. Hence the latter is of the order of 10,000 times smaller than the former. The density of the nucleus is approximately 10^{14} g per cm^3. The forces that hold the nuclear particles are very different from those met with outside the nucleus.

More than 30 elementary particles and antiparticles (like positron, antiproton, etc.) other than protons, neutrons and electrons have been described but their involvement in the constitution of the nucleus is not yet completely clear.

Table 1.1: Properties of elementary particles

Name	Symbol	Mass			Charge*
		g	amu**	Electron mass	
Electron	e^-	$9.106 x 10^{-28}$	0.000549	1	-1
Proton	p	$1.6730 x 10^{-24}$	1.007593	1,836	+1
Neutron	n	$1.6752 x 10^{-24}$	1.008982	1,836	0

* One electron charge = $1.6 x 10^{-20}$ electromagnetic units (amu)
= 1.6×10^{-19} practical units (coulombs)
= 4.8×10^{-10} electrostatic units (esu)

** amu = Atomic mass unit: It is defined as 1/12 of the mass of carbon-12,
(1.66×10^{-24} grams).
1 amu = 931 MeV.

Electron volt : is defined as the energy acquired by an electron in falling through a potential of one volt. Since the electron has a very small mass this unit is exceedingly small in magnitude (= 3.95×10^{-20} calories)

$$1 \text{ eV} = 1.6 \times 10^{-12} \text{ egs.}$$

$$1 \text{ erg} = 6.25 \times 10^{11} \text{ eV} \quad \text{......... (1)}$$

$$1\frac{\text{erg}}{\text{Sec}} = 1 \times 10^{-7} \text{ Walt} \quad \text{......... (2)}$$

1.3 Discovery of the Electron

Then in 1897, **Thompson** discovered the first component part of the atom: **the electron,** a particle with a negative electric charge. In 1904, he proposed an initial model of an atom, since nicknamed "Thompson's pudding".

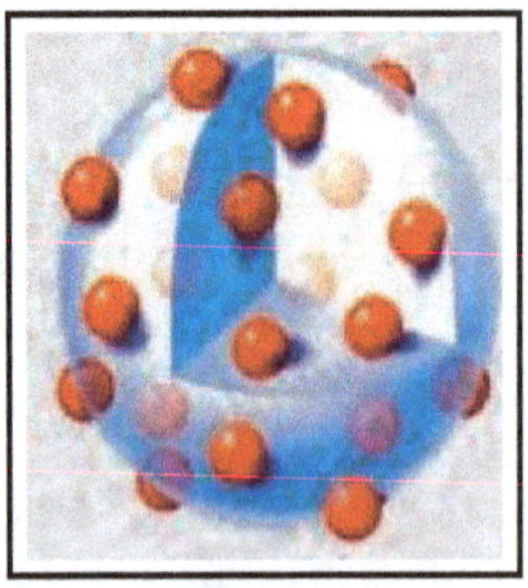

He imagined the atom as a sphere full of an electrically positive substance mixed with negative electron "like the raisins in a cake".

□□□

Chapter 2

Isotopes and Radioisotopes

2.1 Isotopes

In 1913, J.J. Thomson and Aston introduced atmospheric neon into the positive ray apparatus and got two parabolas corresponding to masses 20 and 22, with the intensities of the tracks corresponding to the ratio of 9:1, irrespective of the condition of discharge and pressure. This indicated that neon existed in two forms having atomic weights 20 and 22, and at the same time it could not be distinguished chemically. These two types are known as isotopes of neon (iso = same, topas = place in the periodic table). When the average atomic weight of the ordinary neon was calculated based on the abundance of these two forms in the ordinary neon, it was found to be 20.2 which agreed very well with those determined by the chemical methods. Thus the existence of isotopes of a given element was discovered.

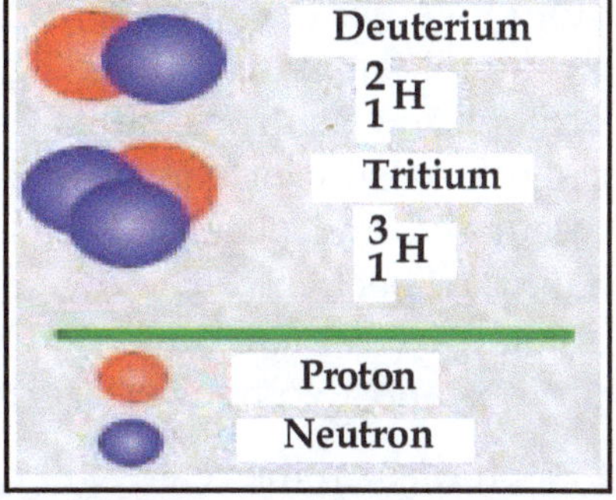

Fig. 2.1: The isotopes of Hydrogen

Formerly isotopes were defined as forms of elements having identical chemical properties but with different atomic weights, the same atomic number and occupying the same position in the periodic classification. Later this definition was slightly modified and now isotopes are defined as nuclides with the same number of protons but differing in the number of neutrons, for example, ordinary hydrogen (also called protium ; proton =1 and neutron =0), deuterium (heavy hydrogen ; proton =1 and neutron = 1), and tritum super heavy hydrogen (proton =1 and neutron =2). Investigations using mass spectrographs have revealed more than 280 isotopes which are stable artificially radioactive isotopes.

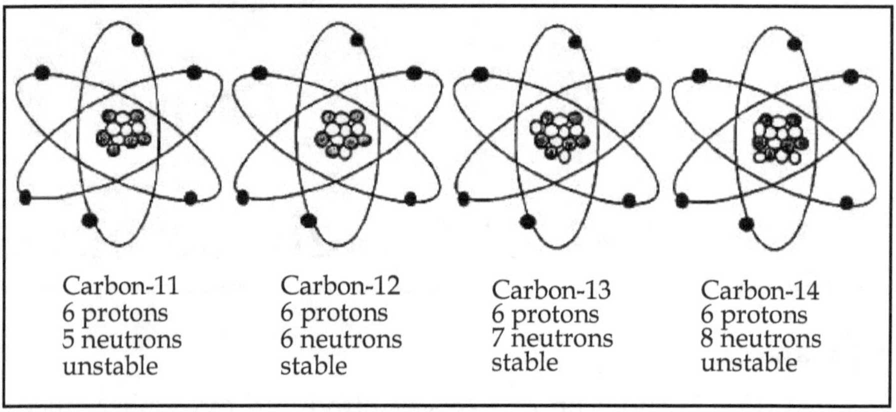

Fig. 2.2: The Isotopes of Carbon

Isotopes generally obey the whole number rule if the masses are calculated with the mass of neutral oxygen taken as 16. The concept of fractional atomic weights has been explained as due to the elements being a mixture of different isotopes. Meanwhile studies by Aston revealed that there is a slight deviation from the whole number rule. If M be the actual mass (weight) of an isotope and A the nearest integer (mass number), then (M-A) is called the mass defect and $\frac{M-A}{A}$ the packing fraction. The mass defect is said to represent the mass that is converted to energy and this is utilized in holding the nucleus of an atom together and represents the stability of the nucleus. Even when the atomic mass unit based on of carbon-12 are used for the mass of protons and neutrons, still a deviation of the observed mass from the calculated sum of constituent masses was found. For example, the calculated and observed masses for helium were 4.033 and 4.0026 a.m.u. respectively. The deviation was found to be equivalent to the amount 0.0304 a.m.u. (1%) of energy (28.2 MeV) released when two protons and two neutrons form a helium nucleus.

Studies on the relative abundance of isotopes of various elements showed that:

1. Elements with even atomic numbers have more isotopes than those with odd atomic number and
2. Isotopes with mass numbers in multiples of four are particularly abundant.

The proportion of each of the isotopes of an element is called the isotopic abundance and expresses the relative amount of the isotopes as percentage. For example, the relative abundances of the stable hydrogen are 99.985 per cent (^{1}H) and 0.015 per cent (^{2}H).

In addition to the nuclides differing only in the number of nuclear neutrons (isotopes), there are nuclides having the same number of neutrons. These are called isotones. Similarly, there are nuclides with the same atomic weight but with different numbers of protons and neutrons. These nuclides are called isobars. For example, has the same mass number 31 but had entirely different chemical properties since they have different atomic numbers. Among the isobars, the stable isobars are found mostly in pairs. No stable isobars and the charge number differ by two units for the members of any pair.

There are nuclides with the same A and Z but differing only in the amount of energy and in the radiation properties. These substances are called nuclear isomers and the corresponding radioactive decay process is called isomeric transition (I.T.).

Table 2.1: Comparision of isotopes, isotones, isobars and isomers

Term	Atomic number (Z)	Neutron number (N)	Mass number (A)	Examples
Isotopes	Same	Different	Different	$^{35}_{17}Cl_{18}$, $^{36}_{17}Cl_{19}$, $^{37}_{17}Cl_{20}$
Isotones	Different	Same	Different	$^{38}_{18}Ar_{20}$, $^{39}_{19}K_{20}$, $^{40}_{20}Ca_{20}$
Isobars	Different	Different	Same	$^{31}_{15}P_{16}$, $^{31}_{16}S_{15}$
Isomers	Same	Same	Same	^{46}Sc, ^{46m}Sc.

2.2 Radio Isotopes

In a stable element, the number of protons and the number of neutrons are equal. When the number of neutrons does not equal the number of protons the atom is called an isotope. Some isotopes have a stable nucleus meaning that the imbalance in protons and neutrons is not enough to cause the nucleus to change. However, in an unstable nucleus, this is not the case and the nucleus changes to attempt to become balanced. As the unstable nucleus changes, it gives off radiation and is said to be radioactive. Radioactive isotopes are often called radioisotopes.

Radioisotopes are unstable, that is, they undergo spontaneous disintegration and give off atomic particle, as a steam of radiation, which can be of different types. The emission of atomic particles can be visualized as flashes of invisible Light other atomic particles given off.

2.2.1 *Alpha (α) Rays (He particles)*

The equation would look like this:-

$$^{241}_{95}Am \rightarrow ^{237}_{93}Np + ^{4}_{2}\alpha$$

$$^{210}_{84}Po \rightarrow ^{206}_{82}Pb + ^{4}_{2}He$$

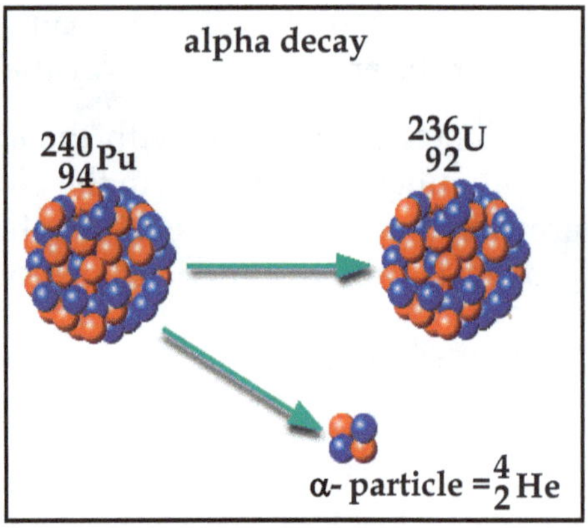

Fig. 2.3: Alpha decay

The alpha particles are generally emitted by the heavy nuclides. They are monoenergetic in nature for each one of the elements, and considering the various elements, it is found to vary from a fraction to 8 MeV. This energy may be spent in electrons. The energy is roughly approximately to R =E (R = range in air in cm; E=energy in ev). These rays have short ranges in air and extremely short ranges in liquids and solids. The range of α ray with 2.0 MeV is 1.05, 0.006 and 0.005 cm in air, aluminum and copper respectively. The range or distance of penetration into matter depends on the initial energy, the particle and the density of the absorber. Alphas lose energy continuously in their path and as the energy is dissipated along the path, the velocity also decreases. Because of the double positive charge, alphas are powerful ionizes and due to their mass they pass through a gas in a straight line. As the alpha emission from a nucleus represents a removal or loss of two protons and two neutrons or a total of four mass units, the atomic weight A is reduced by 4 and the atomic number Z by 2 (Fig-2.3).

2.2.2 *Beta (β) Rays*

Positron ($^{+}\beta$) $\quad {}^{18}_{9}F \rightarrow {}^{18}_{8}O + {}^{0}_{+1}\beta + \nu$

Electron ($^{-}\beta$) $\quad {}^{32}_{15}P \rightarrow {}^{32}_{16}S + {}^{0}_{-1}\beta + \nu$

The beta rays are electrons of nuclear origin and are either positive or negative in nature. The nucleus does not contain free electrons. Hence, the electrons which are emitted as β-rays by the radioactive elements in a proton and electron. Pauli suggested that this process has to be accompanied by the emission of another particle called **Neutrino**. This fundamental particle was "invented" in 1930 by Wolfgang Pauli (him again!) to explain the mechanism of Beta radioactivity: a neutron transforms into a proton by emitting an electron (beta ray) and another mysterious particle that the Italian physicist Enrico Fermi named in 1933 "neutrino"; which in Italian means "little neutron" assumed to be electrically neutral and having a very small rest mass, smaller than that of the electron. In the β emission, the proton remains in the nucleus but the electrons and neutrino are ejected. These results in negligible change in atomic weight in elements but their atomic numbers increase by one. For example, thorium ${}^{230}_{90}Th$ decays to protactinium ${}^{230}_{91}Pa$ with the emission of a β-particle. This β decay results in the displacement of the element by one number in the periodic table. This is called the **"Law of Displacement"**

Neutron → Proton + Electron + Neutrino

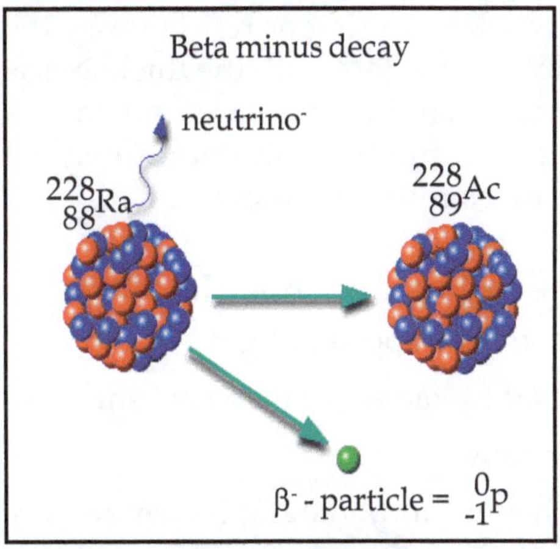

Fig. 2.4: Beta minus (B^{-}) decay

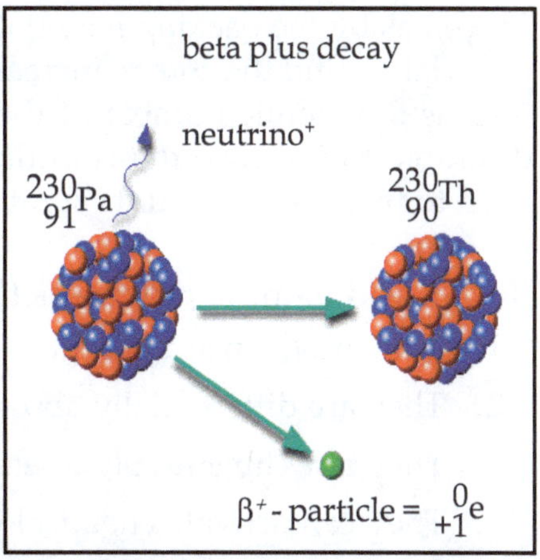

Fig. 2.5: Beta plus (B^{+}) decay

2.2.3 Ionization and ion pairs

When a charged particle passes close to an atom, electrostatic forces of attraction or repulsion operate between it and the orbital electrons resulting in the excitation of electrons or removal of electrons from the atoms. If the electrons are moved, the resulting negative electron and the positively charged ion are called **Ion Pair**. Each radiation particle gives up a portion of its energy by the formation of a long succession of ion pairs. The number of ion pairs produced per cm of the path is known as specific ionization. This value for α, β and Y are 40,000, 50 and 1 respectively. These rays come off with all energies below a certain maximum value, for example ^{32}P = 1.7 MeV; ^{14}C=0.16 MeV. The velocities of betas vary from 0 to 2.9×10^{10} cm/sec nearly equal to that of light. As in the case of alphas, their range into matter is dependent on the initial energy of the particles and the thickness and density of the absorber. But in the case of β-particles, the range is not so well defined due to its zig zag path (scattering) and self absorption. The betas coming out of a solid substance, particularly those with energies less than 0.4 MeV, will be partly stopped by the material itself if the material is of large thickness. Self absorption becomes important when the sample thickness reaches $1/20^{th}$ of the range of betas. The absorption of betas depends upon the number of electrons in its path, and for low atomic weight material, it is a function of distance traveled into the matter and its density. The slope of the curve relating the long of the activity and absorber thickness in known as **Mass Absorption Coefficient** and is characteristic of each isotope. This is used for the identification of the radioisotopes and their radioactive purity. Betas also undergo certain amount of reflection from the sides of the container as well as by the backing material. Back scattering or reflection by the material behind the source increases to an extent with the thickness as well as the atomic number of the material. In some cases, when it is desirable to increase the counting rate, the back scattering may be used to increase the counting rate by as much as 30 per cent.

X-rays and gamma rays have the following properties

1. They travel in straight lines with the speed of light.
2. They are differentially absorbed by materials they pass through.
3. They may liberate photo electrons.
4. They can activate chemical changes in photographic emulsions.
5. They may cause fluorescence.

Table 2.2: Different properties of radiations

Name	Symbols	Identity	RC*	RM*	Penetrating Power	Interaction with Charged Plates	Hazards
alpha	α, ${}^{4}_{2}He$	helium nucleus	2+	4	low - stopped by a sheet of paper	attracted to negative plate, deflected by positive plate	harmful if ingested
beta	β, ${}^{0}_{-1}e$	electron	1-	$\frac{1}{2000}$	moderate - pass through paper and ½mm aluminium, stopped by ½mm lead	attracted to positive plate, deflected by negative plate	skin burns, harmful if ingested particularly iodine-131 in thyroid & strontium-90 in bones
gamma	γ	electromagnetic radiation	0	0	high - only stopped by several centimetres of lead or many centimetres of concrete	unaffected	most dangerous as these are the most penetrating, as a consequence, gamma rays can be used to sterilize materials & destroy bacteria in food

RC* = Relative Charge RM* = Relative Mass

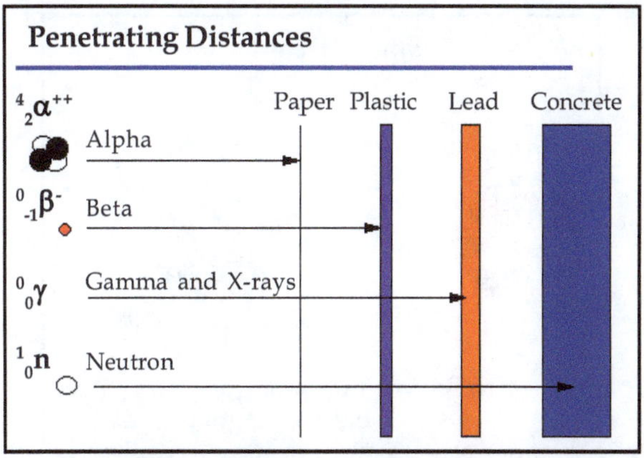

Fig. 2.6: Properties of α, β and γ rays.

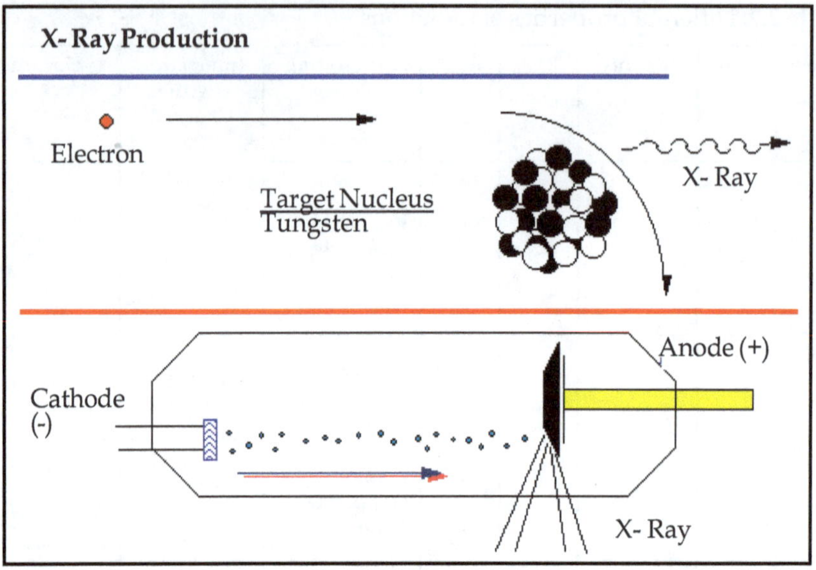

2.2.4 *Gamma (Y) Rays*

$$^{22}_{11}Na \rightarrow ^{22}_{10}Ne + ^{0}_{+1}\beta + v + ?$$

The gamma rays are electromagnetic in nature with shorter wave lengths in the range of 10^{-3} to 10 A^0. These rays travel with the velocity of light (3×10^{10} cm/sec). The electromagnetic radiations originating from the atomic nuclei are termed gamma radiations while those originating outside the atomic nuclei are termed X-rays. Gamma rays are produced when an unstable atomic nucleus releases energy to gain stability through the transition of one nuclear isomer to another or from electron capture (Fig. 2.7).

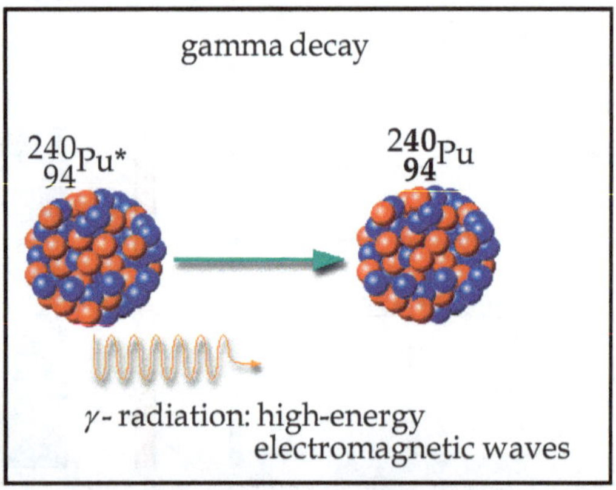

Fig. 2.7: Gamma decay

1. *Compton effect:* When gamma photons of higher energy (0.5 to 1 MeV) approach a molecule, the photons give up only a part of its energy in ejecting an orbital electron and become photons of lesser energy. These again interact with other molecules and this process goes on.
2. *Pair production:* Gamma photons are converted into an electron and position in the region of strong electromagnetic field such as that surrounding a nucleus.

2.3 The Discovery of X-rays

Roentgen

In late 1895, a German physicist, W. C. Roentgen was working with a cathode ray tube in his laboratory. He was working with tubes similar to our fluorescent light bulbs. He evacuated the tube of all air, filled it with a special gas, and passed a high electric voltage through it. When he did this, the tube would produce a fluorescent glow. Roentgen shielded the tube with heavy black paper, and found that a green colored fluorescent light could be seen coming from a screen setting a few feet away from the tube. He realized that he had produced a previously unknown "invisible light," or ray, that was being emitted from the tube; a ray that was capable of passing through the heavy paper covering the tube. Through additional experiments, he also found that the new ray would pass through most substances casting shadows of solid objects on pieces of film. He named the new ray X-ray, because in mathematics "X" is used to indicated the unknown quantity.

In his discovery Roentgen found that the X-ray would pass through the tissue of humans leaving the bones and metals visible. One of Roentgen's first experiments late in 1895 was a film of his wife Bertha's hand with a ring on her finger. The news of Roentgen's discovery spread quickly throughout the world. Scientists everywhere could duplicate his experiment because the cathode tube was very well known during this period. In early 1896, X-rays were being utilized clinically in the United States for such things as bone fractures and gun shot wounds.

2.4 The discovery of Radioactive Elements

Subsequent to Roentgen's discovery of x-rays, in 1896 a French scientist Henri Becquerel was experimenting with a uranium compound. While investigating the properties of fluorescent minerals, it was Becquerel who discovered that certain types of atoms disintegrate by themselves. When working on the principles of fluorescence, he utilized photographic film to record fluorescence of various minerals when exposed to sunlight.

One of the minerals Becquerel worked with was a uranium compound. The experiment normally consisted of wrapping some photographic film in light proof paper, placing a piece of fluorescent uranium on top of the film, and leaving them in the sun. One day, after preparing the experiment, it was too cloudy to expose his samples to direct sunlight, so he stored the uranium compound and the film in a drawer. A couple of days later, he decided to develop this film anyway, and discovered an image of the uranium sample on the film. Becquerel questioned what would have caused this. He knew he had wrapped the film tightly in light proof paper, so the image was not due to stray light. In addition, he noticed that only the film that was in the drawer with the uranium compound had an image on it. Becquerel concluded that the uranium compound gave off something invisible that could penetrate heavy paper and affect photographic film. Becquerel continued to test many samples of compounds and determined that the source of the invisible something was the element uranium. This invisible something was named **radiation**, and it was determined that an element that gives off radiation is a **radioactive element**. Today, we know uranium as one of the radioactive elements. For his discovery of radioactivity, Becquerel was awarded the 1903 Nobel Prize for physics.

Who are the Curies?

Other scientists hard at work discovering radioactive elements were Polish scientist Marie Curie and her husband Pierre Curie, a French scientist. While working in France at the time of Becquerel's discovery, they became very interested in his work. They too suspected that a uranium ore, known as pitchblende, contained other radioactive elements. The Curies started looking for these other elements, and in 1898 they discovered another

Marie Curie

Pierre Curie

radioactive element in pitchblende. They named it 'polonium' in honor of Marie Curie's native homeland. Later that same year, the Curie's discovered another radioactive element which they named radium. Both polonium and radium were more radioactive than uranium.

For their work on radioactivity, the Curies were awarded the 1903 Nobel Prize in physics. In 1910, Marie Curie was awarded the Nobel Prize in chemistry for her discoveries of radium and polonium, thus becoming the first person to receive two Nobel Prizes. Since these discoveries, many other radioactive elements have been discovered or produced. Today many artificial radioactive materials are produced and put to use in various ways ranging from medical to industrial.

❑❑❑

Chapter 3

Radioactivity

The discovery of radioactivity, in general, actually came about on a few different fonts. First was the discovery of x-ray radiation that was artificially generated in a laboratory, followed by the discovery of several elements that naturally emit radiation when the nucleus of the atoms disintegrates or decay. These are the elements that today are called **radioactive elements** and are said to have unstable nuclei.

Table 3.1: Year wise discovery of radiation

Rays	Discoverer	Year
X – rays	Roentgen	1896
α - rays	Becquerel	1899
β - rays	Rutherford	1899
γ - rays	Villard P.	1900

Just before the turn of the century, in the mid to late 1890's, several scientists were working with cathode ray tubes investigating properties of fluorescent minerals. Fluorescent minerals are certain minerals that glow when exposed to sunlight. At the same time, other scientists were busy gathering evidence on the theory that the atom could be divided into even smaller subatomic particles. Some of this new evidence showed that certain types of atoms disintegrate by themselves.

Many of the nuclides have stable nuclei while others are unstable or radioactive. All isotopes of elements above bismuth (Z=83) in the periodic table are naturally unstable and a few of the lighter elements have one or more unstable isotopes. "The spontaneous decay of the nuclei of the unstable isotopes through the emission of α, β and γ rays has been termed radioactivity". Radioactivity can not be altered by physical means such as temperature, pressure, or chemical state. The nuclear decay is a random process and is a manifestation of subatomic changes. Every radioactive substance decays at a constant rate, which is characteristic of each individual substance. No two radioactive isotopes decay at the same rate. The rate of decay, which is called decay constant, is the fraction of atoms decaying per infinitely small time and can be represented by

$$N=N_0e^{-\lambda t} \text{ or } \log N_0-\log N=0.43\ \lambda t$$

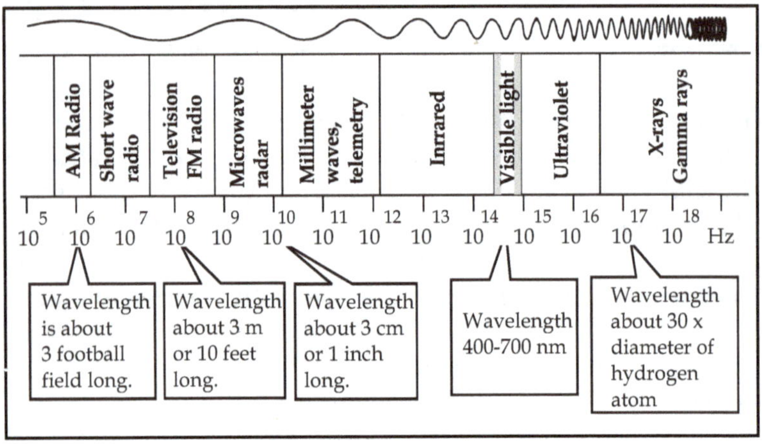

Fig. 3.1: The Electromagnetic Spectrum of radiation

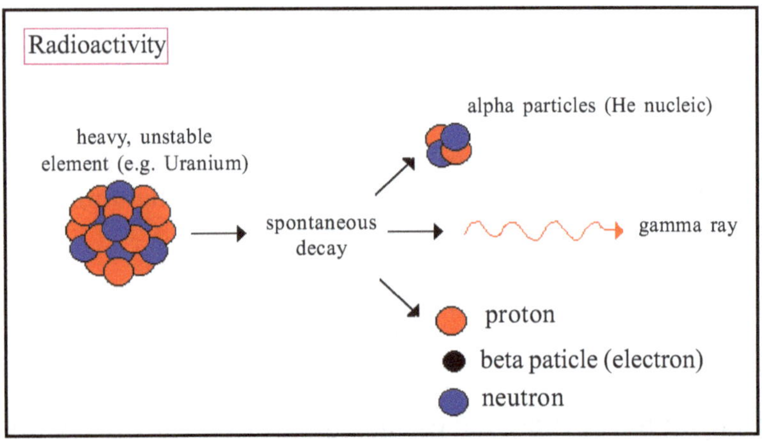

Fig. 3.2: Radioactivity

where N_0 and N represent the activity or the total number of atoms present at zero time and time t respectively, and ë = decay constant. As could be seen from the above equation, the decay constant is characteristic of the element and is not affected by change in environmental factors such as pressure, temperature and chemical state of the element.

3.1 Half Life

The radioactive decay is also represented by considering the time interval required for half of the atoms of radioactive elements to decay or change to another form (daughter nucleus). This interval is known as physical or radioactive half life and is represented by

$T_{1/2} = \frac{0.693}{\lambda}$ *where* $T_{1/2}$ = half-life and λ = decay constant.

The half lives of the various elements differ widely. Half-lives for various radioisotopes can range from a few microseconds to billions of years. See the table below for a list of radioisotopes and each of unique their half-lives.

Table 3.2: Half life of radioactive element

Radioisotope	Half-life
Polonium-215	0.0018 seconds
ismuth-212	60.5 seconds
Sodium-24	15 hours
Iodine-131	8.07 days
Cobalt-60	5.26 years
Radium-226	1600 years
Uranium-238	4.5 billion years
Carbon-14	5760 years
Chlorine-36	$3x10^5$ years

Let's look closely at how the half-life affects an isotope. Suppose you have 10 grams of Barium-139. It has a half-life of 86 minutes. After 86 minutes, half of the atoms in the sample would have decayed into another element, Lanthanum-139. Therefore, after one half-life, you would have 5 grams of Barium-139, and 5 grams of Lanthanum-139. After another 86 minutes, half of the 5 grams of Barium-139 would decay into Lanthanum-139; you would now have 2.5 grams of Barium-139 and 7.5 grams of Lanthanum-139.

The picture below shows how long it takes Phosphrous-32 to decay into Sulphur-32.

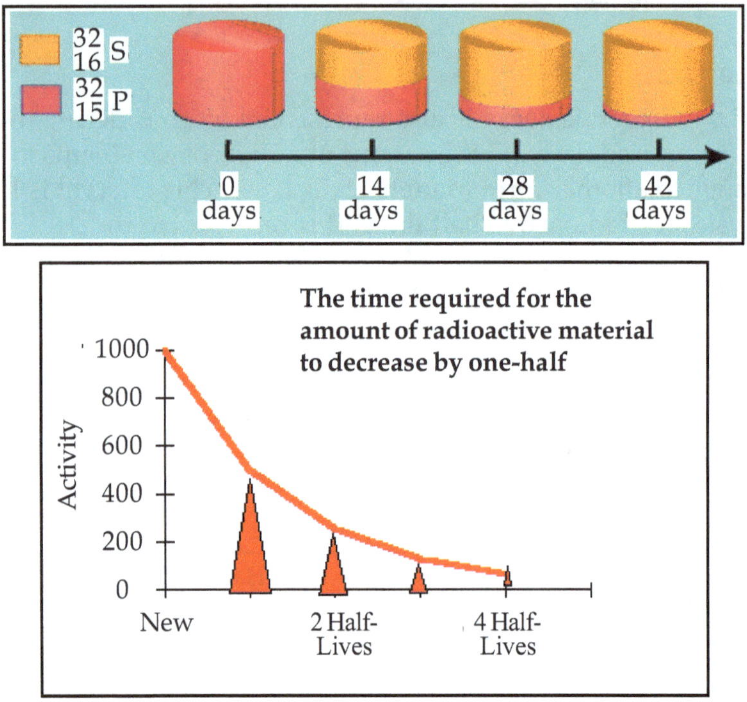

Fig. 3.3: Activity changing in half life of elements with time

3.2 Types of Radioactivity

3.2.1 Natural radioactivity

At the time of earth's creation, probably many radioisotopes were formed but these are not present in natural state today as they are unstable and have decayed to stable forms. The originally formed radioisotopes which still appear today are nuclides that either decay very slowly or are members of a decay series (unstable nuclides which decay to other unstable nuclides and then reach stability only after a dozen or more steps).

Several originally formed natural radioisotopes of the lower atomic number; out of witch only potassium (^{40}K) is biologically important as it is a common constituent of all living tissues. A few radioisotopes like tritium (^{3}H), carbon (^{14}C) are being produced in the earth's upper atmosphere owing to the cosmic bombardment of the atmospheric atoms, and these radioisotopes are readily incorporated into the living organisms, leaving a definite concentration following death. This makes radiocarbon dating possible. In addition to the above, a few of the lighter elements, rubidium (^{87}Rb), samarium (^{147}Sm), lutetium (^{176}Lu), rhenium (^{187}Re) and perhaps one or two other elements possess feeble radioactive properties in their normal state.

The phenomenon of radioactivity was first discovered by French Scientist Henri Becquerel (1906). In 1903, Becquerel and the curies shared Nobel price in physics for their pioneering in radioactivity.

After the identification of polonium and radium by the Curies, a number of new radioactive elements like actinium, uranium and thorium were discovered during the last decade of the nineteenth century and the first quarter of the twentinth century. At present 42 radioactive species of high atomic weights are known to exist in nature. These 42 naturally occurring radioisotopes of high atomic weights, which occur at the end of the periodic table, fall into three distinct series, namely thorium series (4n), uranium series (4n+2) and actinium series (4n+3). The thorium and uranium series are named after the longest lived precursor-thorium and uranium with half lives of 1.39×10^{10} and 4.51×10^{9} years respectively. The parents of these elements undoubtedly had shorter lives and consequently no longer exist in any detectable amounts. The parent of the actinium series is not actinium as was originally thought, but a rather much shorter lived element, sometimes referred to as actino-uranium, with a half life of 7.1×10^{8} years.

In each of the series, there is at least one branches disintegration, as certain elements such as thorium, disintegrate in two ways- one, accompanied by the emission of an α-particle, and the other by a β-particle occur in definite proportion in every case. For example, 33.7 per cent of thorium C atoms give off alpha-particles to form thorium C′ whereas 66.3 per cent emit beta-particles and get converted into thorium C″. Upon subsequent disintegration, both thorium C′ and thorium C″ yield the same product, namely thorium D. Each of the natural radioactive series has a number of collateral members, which have been made artificially and do not exist in nature.

The mass number of thorium (whole number nearest to the atomic weight) is 232, which is 4 x 58. An alpha -particle disintegration results in a decrease of 4 in the mass number while there is no appreciable change in the beta-particles emission. Hence the mass numbers of all the members of the thorium (4n) series can be represented by 4n where n is the whole number varying from 58 to 52. Similarly the mass numbers of the uranium series (4n+2) and the actinium series (4n+3) could be represented by 4n+2 and 4n+3 respectively. In view of the occurrence of the three natural radioactive series with 4n, 4n+2 and 4n+3. Russell, in 1923, suggested the possibility of another series with mass numbers 4n+1. In 1940 Turner proposed a series called neptunium series after the longest-lived member neptunium. The stable end product of the neptunium series is ordinary bismuth of mass number 209 whereas the non-radioactive end products of thorium, uranium and actinium, are only lead of 2.2 x 10^6 years. It is possible that all the members of the neptunium series might have decayed completely and become the non-radioactive element bismuth.

3.2.2 Artificial radioactivity

During the same (1923-34) period when these developments were taking place, Irene Joliot Curie and F. Joliot were studying the effect of bombarding the alpha-particles from the naturally occurring radio element polonium on the nuclei of certain light elements like boron, magnesium and aluminum. They found that there was an emission of neutrons and positrons in addition to protons due to the bombardment with alpha-particles.

$$^{27}_{13}\text{Al} + ^{4}_{2}\text{He} \rightarrow ^{30}_{15}\text{P} + ^{1}_{0}\text{n}$$

Though the formation of protons and neutrons ceased with the removal of the source (α-particles), positron was found to be emitted continuously. The ^{30}P isotope of phosphorus was found to decay.

$$^{30}_{15}\text{P} \rightarrow ^{30}_{14}\text{Si} + ^{0}_{+1}\text{e}$$

Similar transmutations were found to occur with boron and magnesium, the product being radioactive nitrogen ^{13}N and silicon ^{27}Si, respectively.

With the discovery of this phenomenon which was named artificial radioactivity by the Curies, it was decided to add the prefix radio to the names of such unstable isotopes. As many elements exist in several different radioactive forms, the radio nuclides are generally represented by giving the half life and mass number in addition to the chemical symbol of the element, for example, 10 min. ^{13}N. The radioactivity of the substance obtained by nuclear bombardment decays in an exponential manner just like other naturally occurring radio elements.

With this work, a large number of studies were started and more than 1,000 different radioactive isotopes have been prepared since 1934. Though the first artificial radioisotope identified decayed with the emission, of positron, later studies have shown that this could happen with the emission of α- and β-particles. Prior to 1947, the known positron emitting species were restricted to elements of moderately low atomic number but nuclear spoliation has led to the isolation of several radioactive isotopes from fairly high atomic number.

Among the stable nuclides numbering more than 270 that exist in nature, the ratio of neutrons to protons in the nucleus falls within a somewhat limited range and this varies with the atomic number. When the neutron to proton ratio is larger than the stability range, the isotope becomes unstable and tries to acquire stability by replacing a neutron by a proton with the removal of a negative electron as β-particles.

$$\text{N: P High } {}^{32}_{17}\text{P} \xrightarrow{e^-} \left(\text{P}^+\right) \rightarrow {}^{32}_{18}\text{S} + \beta^- \, (electron)(Neutron\, decay)$$

$$_{17}\text{P}_{15\eta} \rightarrow {}_{16}\text{P}_{16\eta} + \beta^- \, (\text{electron})$$

Such nuclei are negative, β^- active and have mass numbers larger than those of the stable isotopes (^{14}C, ^{16}N, ^{19}O, ^{28}Al, ^{38}Cl, ^{59}Fe and ^{83}Br).

When the neutron to proton ratio of nuclides lies below the range of stability, the neutron deficient nucleus will tend to change in such a manner as to replace a proton by a neutron by any one of the following processes.

1. Emission of positron:

$$^{26}_{13}\text{Al} \longrightarrow {}^{26}_{12}\text{Mg} + \beta^+$$

2. Expulsion of an alpha particle:

$$^{232}_{90}\text{Th} \longrightarrow {}^{228}_{88}\text{Ra} + {}^{4}_{2}\text{He}$$

$$^{210}_{84}\text{P}_\text{o} \longrightarrow {}^{206}_{82}\text{Po} + {}^{4}_{2}\text{He}$$

3. Capture of an orbital electron:

$$_{24}^{54}Mn + \alpha \longrightarrow {}_{24}^{54}Cr + p + \alpha^{-}$$

$$_{17}^{55}Fe + {}^{0}e_{1} \longrightarrow {}_{25}^{55}Mn$$

Positron activity is commonly observed with isotopes having mass numbers smaller than the stable values, for example, ^{11}C, ^{13}N, ^{15}O, ^{26}Al, ^{32}Cl, ^{53}Fe and ^{78}Br. Nuclear reactions of the (p, n), (p, n), (p, α), (α, n) (n, 2n), (γ , 2n) types occur and decrease the relative proportions of neutrons, making them frequently positron active.

In high energy spoliation reactions it frequently happens that the particle in the nucleus which splits off first is the neutron, leaving neutron deficient nuclei with atomic numbers not very different from those of the target element. Such nuclei are radioactive and decay, as is to be expected, by the emission of positron. Although neutrons are also emitted in nuclear fission, the number is relatively small. The product nuclei still have too many neutrons for stability and are consequently negative β-emitters.

In a few cases, radioactive decay occurs with the emission of both positron and negative electrons. This is a type of branched disintegration in which part of the nuclei decay in one manner and the other part in a different way. The mass numbers of stable copper isotopes are 63 and 65, so that ^{64}Cu has a neutron to proton ratio which lies between the values of the two stable species.

When a radio active element emits a negative, β-particle, the product has the same mass number as the parent but its atomic number is greater by one unit. When a positron is emitted, the mass number is still unchanged but the atomic number of the product is now one unit less than that of the parent because the positron emission is associated with the replacement of proton by a neutron. The change clearly leaves the total mass unaffected but decreases the number of protons and hence the nuclear charge and the atomic number by unity.

Except the naturally occurring Samarium isotope $_{62}Sa^{147}$, the nuclides with atomic number 83 or more were thought to be alpha-emitters for many years. Later work has shown that other atomic numbered elements are also α-emitters.

With the majority of artificial radio nuclides, other than those obtained by nuclear fission and spoliation, the immediate decay product is stable species since the radio isotopes are usually obtained by a simple transmutation wherein a particle enters a stable nucleus and another particle is expelled. In spoliation, however, the splitting off of several nucleons may leave a product that is two or three or possibly more stages removed from stability. Thus, short chains involving two or three successive positron emissions are possible before a stable nuclide is reached.

$$^{71}_{34}\mathrm{Sc} \xrightarrow{\beta^+} {}^{71}_{33}\mathrm{As} \xrightarrow{\beta^+} {}^{71}_{32}\mathrm{Ge} \xrightarrow{\beta^+} {}^{71}_{31}\mathrm{Ge}$$

where, $^{71}_{31}Ge$ is a known stable isotope of Gallium.

3.3 Orbital-Electron Capture

Though normally in nuclides with low neutron to proton ratio, positron activity is expected, still another type of decay also occurs, namely the nucleus capturing one of the extra nuclear (orbital) electrons which immediately combines with a proton to form a neutron and a neutrino. The product of this type of radioactivity would have the same mass number as its parent but its atomic number is less by one unit, just as in the case of positron emission. This phenomenon is referred to as decay by orbital electron capture. The electron that is captured is usually from the first quantum level (K-shell) as it is found near the nucleus.

$$^{55}_{26}\mathrm{Fe} \xrightarrow{{}_{-1}e^0} {}^{55}_{25}\mathrm{Mn}$$

Hence this phenomenon is called Orbital Electron Capture. The removal of one electron from the K-shell leaves a vacant position in that shell which is immediately filled by an electron from the higher energy level by emitting the excess energy. Since orbital electron capture must proceed electronic transition and emission of X-rays, the latter will be characteristic of the product nucleus with an atomic number of one unit less than that of the radioactive species. It has a result of orbital electron capture, the product nucleus is left on its ground state, and the change will not be accompanied by the γ-rays. But in most instances, however, the product nucleus is formed in a high energy (excited) state and the excess energy is given off in the form of radiation. Frequently this radiation is internally converted, and the energy of the γ-rays photon is transferred to an orbital electron, which is consequently ejected.

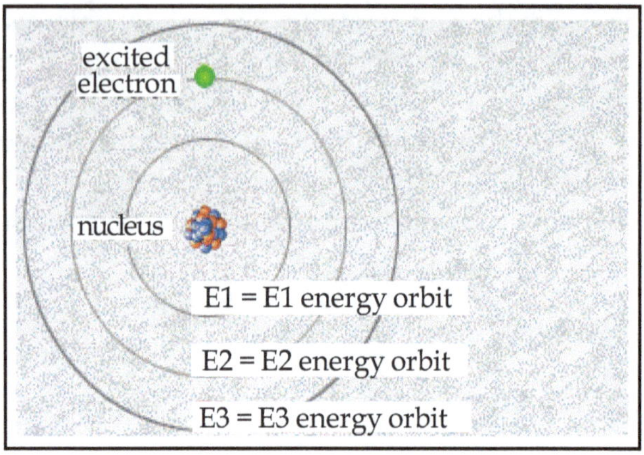

Fig. 3.4: Electron orbit

3.4 Nuclear Stability

We have examined the radioactive decay for various nuclides in some detail, but we have not discussed why some nuclides decay and some are stable. At its deepest level this is a complex and only partly resolved, question. However, various approaches to the problem greatly increase our understanding. Here, we will continue ourselves to certain, empirical criteria that describe general conditions of nuclear stability.

When we plot a graph of "Number of neutrons" against "Number of protons", we find that **stable** elements lie on a "stability line".

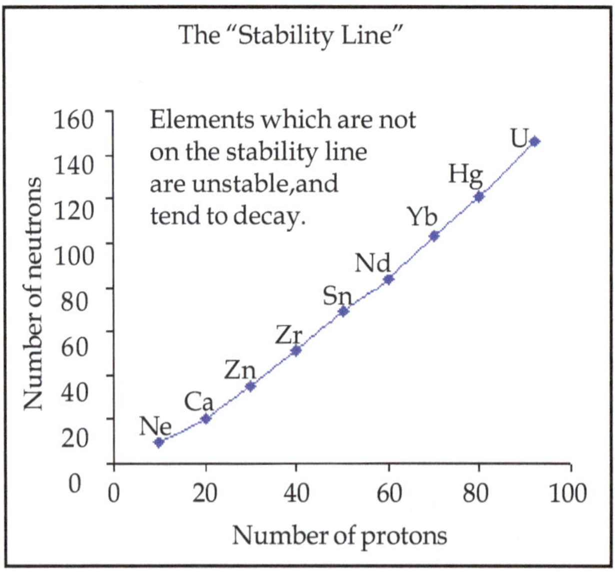

Elements which are not on this line are unstable, and we find that they tend to undergo alpha-decay or beta-decay and get closer to the stability line.

They may take several steps in order to achieve this, thus we observe **decay chains** for most radioactive elements

Table 3.3: Distribution of neutron and proton number of stable nuclides

Proton number	Neutron number	Number of stable nuclides
Even	Even	168
Even	Odd	056
Odd	Even	051
Odd	Odd	004

We summarize nuclear stability criteria this way :

a. *The Upper limit of nuclear size:* There are no stable nuclides whose proton number exceeds 83 or whose neutron number exceeds 126.

b. *Approximate equality of neutron and proton numbers:* For element up to and including phosphorus (Z=15) the neutron number does not exceeds the proton number by more than 2. Only for the nuclides ${}_{1}^{1}H$ and ${}_{2}^{3}He$ do the proton number exceeds the neutron number.

c. *Preference for even numbers of proton & neutron :* About 60% of stable nuclides are even-even; of the remainder, only ^{2}H, ${}_{3}^{6}$Ti ${}_{5}^{10}B$ and ${}_{7}^{15}$N are odd-odd.

❑❑❑

Chapter 4

Isotopes as Tracer

4.1 Selection of Isotopes for Tracer Studies

In radioisotopes we have all the essentials of a powerful research tool, a non-destructive means by which atoms could be followed and detected in the organism. Application mainly depends on the fact that chemical properties of the isotope of an element are essentially identical.

The basic requirement of a tracer is that it be

1. Chemically and physically, virtually equivalent to the substance it represents or displaces and that it, in no appreciable way, affect the system differing from its normal counterpart.
2. Radioactivity could be detected directly and
3. Unaffected by the physical or chemical treatments that are given. Hence, this has been used widely in the investigations of the various problems in biology, chemistry, physics and other fields like agriculture.

The use of tracer techniques dates back to 1911 when G. Hevesy in Austria used the active thorium to demonstrate to his landlady that she was incorporating the remaining of the Sunday pie in the meals served during the subsequent days. Later in 1913, Hevesy and Paneth used radium as a tracer for studying the solubility of various lead salts. In 1923, for the first time in biological studies, he used thorium B (^{212}Pb) to study the uptake of lead by plants. Again it was Hevesy who in

1935, used radioisotope ^{32}P to study the absorption of P by plants from the nutrient solutions. The first non-radioactive isotope that was used was heavy hydrogen (deuterium, ^{2}H) and this was followed by the other stable isotopes like carbon-13, nitrogen-15, and sulphur-34. However, before 1946 the use of isotopes as tracers was limited due to the availability of limited number of stable and radioactive isotopes. The position had changed considerably in the past three decades with the advent of nuclear reactors and also with the discovery of various extraction procedures for fission products.

1. Since the measured radioactivity of the sample is an accurate indicator of the amounts of the radioactive material it contains, the use of isotopes in various fields has increased considerably.
2. Moreover, the radioisotopes could be traced through all paths of complicated chemical reactions by simple measurements and without the use of chemical analysis.
3. Also using the penetrating radiations, measurements can be made even from outside the vessel without disturbing the chemical processes 'inside'. Further, the chemical or element to be followed by this method need not be composed entirely of the radioactive atoms or the stable isotope. If the radioactive and stable atoms are mixed together and in the same chemical forms, any subsequent chemical treatment affects the stable and radioactive isotope in the same manner and hence they will appear in the same proportion in any sample. However, any atom which randomly exchanged position with other atoms in organic combinations cannot be used. So it was of fundamental importance to establish the fact that physical interchange did not occur. When this was proved, it became practical to introduce marked atoms into any biological system which could be sufficiently isolated for study and to follow the normal course of reaction throughout the process.

4.1.1 Choice of isotope

However, before employing the isotopes in the investigations, a choice has to be made between the stable isotopes and radioisotopes, and among the radioisotopes, those having long half life and short half life. For example, in the case of carbon, a choice has to be made between stable isotope ^{13}C (stable isotopes) and the radioactive isotopes ^{11}C (20.5 minutes) and ^{14}C (5760 years). Generally the radioisotopes are preferred in view of their ease of detection, estimation of even small quantities without destroying the organism, capacity to follow their movement, presence of various elements throughout the crop growth etc. The radioisotopes emit different types of radiations like α, β and γ rays. The alphas are generally unsuitable for tracer work because of their lower penetrating power and also due to their inherent counting problems. Generally betas are preferred when the samples are to be removed from the system before counting, since they give a good local concentration and be counted with ordinary counters. Gamma radiations could be counted in situ as well as from outside the system. Further, in view of their high penetrating power, they may cause injury to the organism and hence only small doses of the same could be administered. And the choice of radiations is limited since certain of the elements emit only one type of radiation. In such cases, the tracer techniques have to be suitably modified to make the best use of the available isotopes.

In choosing the radioisotope, attention has to be paid to the specific activity of the element. The radioisotope should be available in sufficient amount and also the organism should consume the particular isotope of the compound in sufficient quantities. Sufficient amounts of the radioisotope should be tagged to the compound, at the same time without causing injury to the organism and in such does that is only the minimum amount of the element to be detected, yet high enough to be detected by a counter. Further, one has to ensure that the entire radioactivity is in the form in which it is to be traced and is firmly bound to the compound or molecule to which it is attached. Also, while interpreting the results, one has to be aware of the possible chemical conversions and transformations in the various pathways and also the possible effects due to dilution in the organs of the organism and also in the various solutions.

As the radioisotopes decay spontaneously, the amount of tracer present will be decreasing gradually and in the course of time may come to undetectable levels. But at the same time, if a high dose of a radioisotope is used for getting detectable amounts at the end of the experiment, the organism may suffer radiation hazards due to the dosage. Normally, the duration of the experiment is adjusted not to exceed 10 half lives of the element. If both short half life and long half life radioisotopes are available as in the case of carbon (^{11}C and ^{14}C), a choice can be made depending upon the requirements of the experiments. In the case of long term experiments in agriculture and other fields, ^{14}C is selected while in the fields of medicine where repeated doses have to be given, short half life radioisotopes like iodine-131 (8.08 days) and carbon-11 (20.5 minutes) are preferred. At the same time, the experimenter has to be aware of the fact that the energy of the beta-particles emitted by a long half life isotope is low, since there is an approximately inverse relationship between the half life of the radioisotope and the energy of the beta particle emitted. But this has been overcome by the modifications of the various techniques. At the same time, short lived isotopes emit high energy β-particles which may produce harmful effects on the experimental organisms and also on the experimenter. Hence such isotopes have to be used in very small amounts much below the tolerance level. All these aspects have to be taken into consideration while making the choice of the radioisotopes.

In cases where radioisotopes of the required long half life are not available, as in the case of nitrogen and oxygen which have radioisotopes of half lives of only 10.1 minutes and 2.1 minutes respectively, one has to resort to the use of stable isotopes. The uses of stable isotopes as tracers were made possible by Urey's method of separating, ^{15}N, ^{13}C, ^{18}O, and ^{34}S in usable quantities. Stable isotopes concentrated above or diluted below their natural abundance level may be used as tracers.

1. The permanency of the stable isotopes,
2. Adaptability to accurate quantitative work,
3. The negligible physiological effects on the living organisms (except for deuterium) and
4. Lack of radiation hazard is distinct advantages they have over radio-isotopes.

But at the same time, use of stable isotopes is restricted due to

(a) The limited availability and
(b) High cost of mass spectrometer or photo spectrometer required for their estimation. Further,
(c) The samples have to be processed before estimation as opposed to the radioisotopes in the case of which the amount could be estimated even without destroying the sample. Still this limited availability of concentrated stable isotopes is offset by the fact that the available ones are of great biological importance. If others are needed, they could also be concentrated and produced.

Further, in the nuclear transformation reactions, all the target atoms do not absorb the neutrons. Only a very small fraction, perhaps one in 100 millions, does. The rest of the atoms remain as ordinary atoms. This inactive material associated with the radioactive isotopes and chemically inseparable from them is known as carrier and the radioactivity per gram of carrier is known as the specific activity. Owing to the changes in the atomic number which take place in the (n, p) or (n, α) reactions, the radioactive material can be separated chemically from the rest of the samples and can be made to have very high specific activity.

However, the source of starting material usually presents a production problem which is very important for the use of radioisotopes. If, for example, ^{32}P is produced from ^{31}P by the ^{31}P (n, γ) ^{32}P reactions, it will be inseparably mixed with the un-reacted ^{31}P. The specific activity is unavoidably low and its use as a radiotracer is limited. If on the other hand ^{32}P is obtained from ^{32}S (n, p) ^{32}P reactions, the product could be readily separated by the chemical means from the un-reacted ^{32}S so that almost all the P in the sample will be radioactive ^{32}P. Such a sample would be known as carrier free, that is, free from any stable isotopes of the same element. Unfortunately for some nuclides, a carrier free stable is impossible owing to the limited choice of the starting material available.

Hence, in planning a tracer experiment, consideration should be given to the

1. Type of experiments
2. The nature of isotopes available
3. The radiations emitted by them
4. The availability of the detecting equipment
5. Half life of the radioisotopes
6. Dose at which they are to be administered without any harmful effects on the organisms
7. Dilution effects
8. Radiation hazards
9. Counting efficiency
10. Purity of the isotopes
11. Specific activity, etc.

In the estimations of the isotopes, it is relatively harmless to maintain high rates for counting than to attempt to maintain low counts as the latter may give a wrong picture and prove to be catastrophic.

In agriculture more than 30 elements have been found to take part in various growth processes either as macronutrients or micronutrients, or have been used for determining the moisture content or for irradiation of the plants or for following the pathway of certain compounds in the plants or for detection of certain contaminants in soil and plants. A list of stable and radioactive isotopes available and those that could be used for tracer studies are furnished in the table 10.1.

4.1.2 Stable Isotopes as Tracers

There are four stable or heavy isotopes of potential interest to researchers in soil and plant studies. These are ^{18}C, ^{2}H, ^{13}C and ^{15}N of these ^{2}H (deuterium) and ^{18}O are used for rather specialized applications, but ^{13}C having gone through a period when it lost ground to the newly developed and than cheaper ^{14}C is now enjoying renewed interest following its increased production at a much lower price ^{15}N is however the heavy isotope of broad application, but ^{13}C, ^{18}O and ^{34}S are readily available.

The radioactive isotopes of nitrogen have too short half lives to be of any general value thus for ^{13}N it is 10 minutes, for ^{16}N seven seconds and for ^{17}N four second. Therefore, although there is a theoretical possibility of using ^{13}N for very short uptake experiments it is clear that the stable ^{15}N isotope is the most suitable isotope for long term studies. Comparatively recently the use of ^{14}N depleted nitrogen compounds has opened up a new approach to studies with nitrogen isotopes.

The stable isotope ^{15}N is used as a tracer in exactly the same way as radioactive isotopes are used except that we identify the tracer atoms by mass differences. However, instead of the specific activity of a sample used in the case of radioisotopes, the term % abundance is used for stable isotopes; the % ^{15}N abundance is the ratio of ^{15}N to ^{15}N + ^{14}N atom. i.e

$$\% \,^{15}\text{N abundance} = \frac{\text{atoms } ^{15}\text{N}}{\text{atms } ^{15}\text{N} + \text{atms } ^{14}\text{N}} \times 100$$

Since the natural environment has on ^{15}N abundance of 0.37% (0.3663%) the amount of ^{15}N in a sample is conveniently expressed as % ^{15}N atoms excess over the natural abundance of 0.37%.

Practical consequence of having to identify ^{15}N by means of mass difference and to define its presence as % abundance is that tracer techniques for stable isotopes are much less sensitive than for radioisotopes, the latter being able to trace at concentration as low as 10^{-11} (1 to 100 billion). However, the never double collector mass than $\pm$ 0.05% has been reported for the micro mass model 60C. This is a considerable advance. But assuming that an older single collector mass spectrometer is capable of determinations with a precision of 0.5 relative % on samples of any ^{15}N abundance down to normal abundance and the normal abundance is 0.37% than the theoretical limiting excess values which can be detected will be roughly 0.002 atom % excess or somewhat less in other words an initial does of nitrogen with 20% atom excess ^{15}N can be diluted only 10,000 times if the ^{15}N is still remain detectable (not quantitatively) in the samples. In practice the analytical error is about 0.01% atom excess ^{15}N, so far quantitative determinations we can not work with dilutions as great as this and final samples for analysis should contain 0.15% atoms excess ^{15}N or more for routine analysis with a single collector.

This greatly decreased sensitivity with increasing dilution inevitably affects the design and practicability of soil-plant experiments with ^{15}N. The one per cent atom excess ^{15}N is adequate for fertilizer experiments where the crop takes up as substantial portion of the applied fertilizer. However, for soils experiment such as studies on turnover processes in the nitrogen cycle, the fractionation and movement of applied nitrogen fertilizer, studies of gaseous losses then much higher ^{15}N atom excess material must be used to overcome the enormous dilution effect such experiments require material with 30-35% N atom excess or over more and materially influence the cost of the work and the size of the experiment that can be conducted.

❑❑❑

Chapter 5

Carbon-14 Dating

Carbon dating uses the half-life of Carbon-14 to find the approximate age of an object that is 40,000 years old or younger.

5.1 What is radiocarbon dating?

Radiocarbon dating is a method of estimating the age of organic material. It was developed right after World War II by Willard F. Libby and coworkers, and it has provided a way to determine the ages of different materials in archeology, geology, geophysics, and other branches of science. Some examples of the types of material that radiocarbon can determine the ages are wood, charcoal, marine and freshwater shell, bone and antler, and peat and organic-bearing sediments. Age determinations can also be obtained from carbonate deposits such as calcite, dissolved carbon dioxide, and carbonates in ocean, lake, and groundwater sources.

5.2 How is Carbon-14 Produced?

Cosmic rays enter the earth's atmosphere in large numbers every day and when one collides with an atom in the atmosphere, it can create a secondary cosmic ray in the form of an energetic neutron. When these energetic neutrons collide with a nitrogen-14 (seven protons, seven neutrons) atom it turns into a carbon-14 atom (six protons, eight neutrons) and a hydrogen atom (one proton, zero neutrons). Since Nitrogen gas makes up about 78 percent of the Earth's air, by volume, a considerable amount of Carbon-14 is produced. The carbon-14 atoms combine with oxygen to form carbon dioxide, which plants absorb naturally and incorporate into plant fibers by photosynthesis. Animals and people take in carbon-14 by eating the plants.

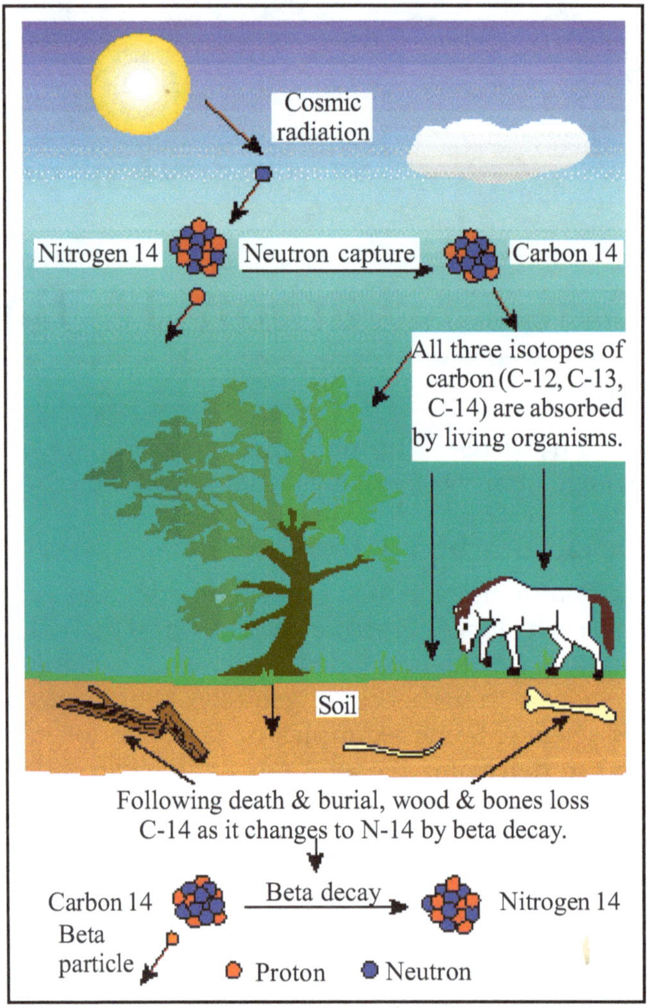

Fig. 5.1: Determination of age fossil through C-14 dating

The ratio of normal carbon (carbon-12) to carbon-14 in the air and in all living things at any given time is nearly constant. Maybe one in a trillion carbon atoms are carbon-14. Both Carbon-12 and Carbon-13 are stable, but Carbon-14 decays by very weak beta decay to nitrogen-14 with a half-life of approximately 5,730 years. After the organism dies it stops taking in new carbon.

5.3 Use of Carbon-14 to Determine the Age of a Fossil

To measure the amount of radiocarbon left in a fossil, scientists burn a small piece to convert it into carbon dioxide gas. Radiation counters are used to detect the electrons given off by decaying carbon-14 as it turns into nitrogen. In order to date the fossil, the amount of carbon-14 is compared to the amount of carbon-12 (the stable form of carbon) to determine how much radiocarbon has decayed. The ratio of carbon-12 to carbon-14 is the same in all living things. However, at the moment of death, the amount of carbon-14 begins to decrease because it is unstable, while the amount of carbon-12 remains constant in the sample. Half of the carbon-14 degrades every 5,730 years as indicated by its half-life. By measuring the ratio of carbon-12 to carbon-14 in the sample and comparing it to the ratio in a living organism, it is possible to determine the age of the fossil.

❑❑❑

Chapter 6

Radioactive Nuclides

Although a number of radioactive isotopes are found in nature, most of the isotopes used as radioactive tracers are produced artificially. Methods for producing radioactive isotopes include nuclear fission and the bombardment of selected materials by neutrons or positively charged particles.

6.1 Production of Radioactive Nuclides

Nuclear Reactors

In a nuclear reactor, a fissionable nuclear fuel such as ^{235}U or ^{239}Pu is placed in a defined geometry within the reactor core. Fast moving neutrons released as the ^{235}U or ^{239}Pu nuclei fission are slowed to thermal energy (0.025 eV) by a low-Z moderator such as water, heavy water, beryllium, or graphite surrounding or mixed intimately with the nuclear fuel. The slowly moving neutrons are absorbed by addition of ^{235}U or ^{239}Pu nuclei, and a chain reaction is maintained. The rate of absorption of slow neutrons by fissionable nuclei determines the fission rate for the reactor and is controlled by placing inert neutron absorbers in the reactor core. Typical neutron absorbers, usually termed control roads include boron and calcium. Heat generated in the reactor is removed by a coolant such as water or one of a variety of gases or liquid materials.

Many uses have been found for nuclear reactors, including the production of power, the production of nuclear fuel, the testing of materials exposed to large amounts of radiation and the production of radioactive isotope. Most of the countries have reactors suitable for producing radioactive isotopes.

Koeberg Nuclear Power Station

Radio nuclide Generators

Radio nuclides generators furnish short lived radioactive isotopes which otherwise might be unavailable because of the decay of the isotopes during transport from remote site. In a generator, radioactive isotope with a relatively long half-life is attached firmly to an ion-exchange column. The isotope decays to a radioactive daughter with a shorter half-life. The daughter is an isotope of a different element and is attached only loosely to the column. Consequently, the daughter may be removed with an effluent such as dilute ammonium hydroxide, hydrochloric acid, or saline water. If the daughter is removed by elution ("milking"), it re-accumulates on column and rapidly attains a state of transit equilibrium with the longer lived parent.

A radio nuclides generator should furnish a product in a chemical from, which requires little preparation before use. Further more; the product should not be contaminated by undesired radioactive or stable isotopes. To reduce difficulties in shipping and shielding, the generator should be ranged and compact. The ion-exchange resin should not decompose during irradiation. A typical generator configuration is diagram in Figure 6.1.

A nuclear power plant produces electrical energy in a way that is animated in Fig. 6.2.

6.2 Isotope Production

Most radioactive isotopes used in biological studies are produced in a nuclear reactor by neutron bombardment of a selected target.

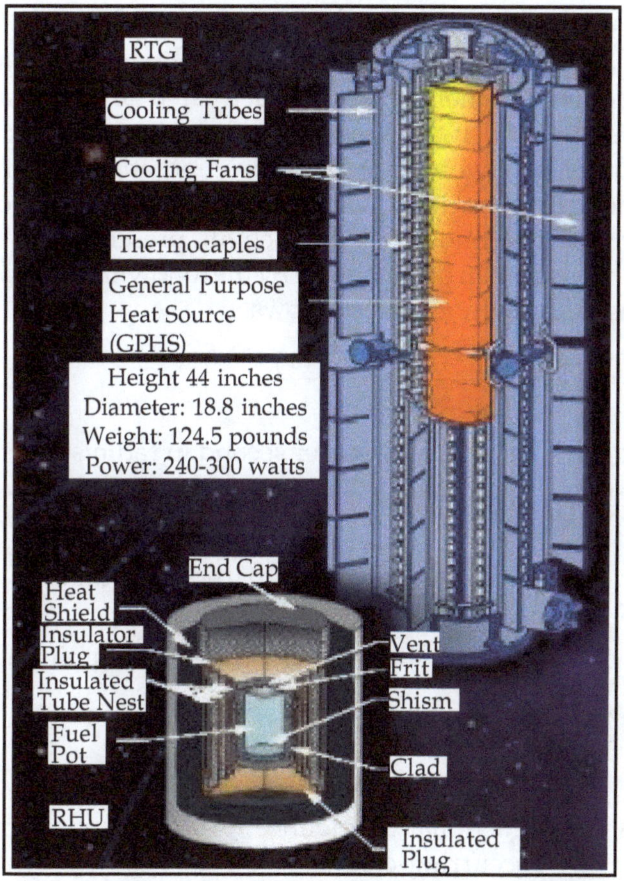

Fig. 6.1: Radionuclide Generators

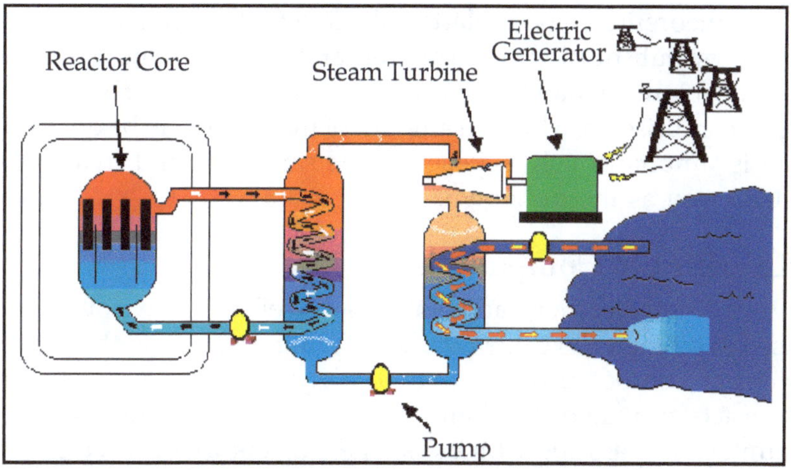

Fig. 6.2: Flow chart of radio nuclide generator

Equation for isotope production

When a sample is exposed to a flux of bombarding particles, the elements contained therein become activated to form radio active species. There are then two opposing processes in operation: the growth of radioactivity, as governed by the flux of radiation and cross section of the element; and the decay, as governed by the half-life of the radio isotope formed. The amount of radioactivity present at any given time during the exposure can be expressed by the following equation, derived from the usual growth and decay lows.

$$A = \frac{G\,(6.02 \times 10^{23})\, f\, \sigma\, (1\text{-}e^{\lambda t})\,(\theta)}{M} \qquad \text{(Eq.6-1)}$$

where, A = disintegrations per second in radioisotope produced.

G = grams of element of natural isotopic composition in sample.

f = flux of bombarding particles in units of particles / cm^2/sec.

σ = cross section of nuclear reaction in units of cm^2/target atom.

λ = radioactive decay constant.

t = time of bombardment.

θ = fractional abundance of target isotope in naturally occurring element.

M = atomic weight of element.

It is theoretically possible to calculate G, the grams of unknown element, by substitution of measured and known values in Eq (6-1). However, it is much more practical to evaluate G by use of a comparison sample, comprised of a known amount of the element to be determined, which is simultaneously exposed to the same flux and treated in the identical way as the unknown.

6.3 Labelling of Compounds

As indicated earlier, radioisotopes as well as stable isotopes could be used in biological work only after tagging or labelling the particular element or compound or animal. The characteristic property of the isotope acts as a tag or label and these are measured either by the use of counters or mass spectrometers. The element that is used as a label or tag is called the tracer element.

Production of Radioisotopes

At BARC: Hot Cell facility for production of radioisotopes

One of the methods of labelling an element is to change its natural isotopic composition. For example, carbon contains 98.9 per cent of carbon-12 and 1.11 per cent of carbon-13. This abundance ratio can be changed either by increasing the percentage of carbon-13 or by adding a radioactive isotope (^{11}C or ^{14}C) that is not found in natural carbon.

In studies where the source of one particular element in the end product has to be found from the two or more possible sources (as in the case of photosynthesis about the source of oxygen-water or carbon dioxide), labelling is achieved by chemical means or by biological synthesis. Labelling of a compound by chemical means requires prior knowledge of the mode of preparation of the compound from its elements or from simple compounds. For example, in acetic acid (CH_3COOH) the carbon of the methyl group (CH_3^-) or the carboxyl group (-COOH) could be labelled by following suitable preparation techniques. Labelling the methyl group (CH_3-) can be done by treating barium carbonate $BaCO_3$ (labelled with C^{14}) with acid. This would release CO_2 (carbon dioxide), and when this is treated with hydrogen under pressure and at high temperatures, CH_3OH (methanol) is obtained. This is converted into CH_3CN (acetonitrile) by treating it with KI first to yield CH_3I, and then with NaCN. On further hydrolysis CH_3CN yields CH_3COOH with the carbon of the methyl group labelled.

$$BaCO_3 \xrightarrow{\text{Acid}} CO_2 \xrightarrow{\text{H}} CH_3OH \xrightarrow{\text{KI}} CH_3I \xrightarrow{\text{NaCN}} CH_3CN \xrightarrow{\text{Hydrolysis}} CH_3COOH$$

In case the reaction is started with unlabelled $BaCO_3$ and ^{14}C labelled NaCN, then acetic acid labelled in the carboxyl group (COOH) will be obtained.

$$BaCO_3 - CO_2 - CH_3OH - CH_3I \overset{\text{NaCN}}{-\,-\,-\,-\,-} CH_3CN - CH_3COOH$$

Similarly, most of the compounds can be prepared adopting suitable labelling procedures. For example, for ^{32}P labelled PCl_3 and $POCl_3$, dialkyl phosphates and dialkyl chlorophosphates can be prepared.

But in certain biologically important compounds like hormones, vitamins, alkaloids, antibiotics, labelling by chemical means will be difficult. In such cases biosynthetic labelling procedures are followed. For example, when the green plants are grown in the presence of labelled CO_2, the carbon of starch produced by the plant would be labelled with ^{14}C. Starch which is a polymer of glucose, on hydrolysis, gives the labelled glucose molecules. The labelled glucose, when combined with fructose having no labelled carbon, will produce sucrose having only the carbon atoms of glucose mostly labelled. Similarly, many of the compounds that are difficult to prepare by chemical means can be prepared by biosynthetic procedures.

Labelling can also be done using stable isotopes. As indicated earlier, the natural composition of the element could be modified by increasing the content of one of the stable isotopes or by adding radioisotopes. Wherever a mass spectrometer is available and also where it is not possible or necessary to label the compounds with radio isotopes for some reason or the other, the compound can be labelled with stable isotopes. Many a time the stable isotope deuterium is used to replace one or more of the associated hydrogen in the compounds. However, in labelling with deuterium, the possibility of isotopic exchange of the stable isotope with other related groups like water has to be always considered and necessary corrections made in the interpretation of the results. Also, while analyzing the data, the effect of increase in density of the compound particularly in low molecular compounds, have to be corrected.

6.4 Double labelling

Sometimes the double labelling technique is employed wherein two radio isotopes of differing elements like (^{14}C - ^{32}P), (^{32}P -^{42}k), (^{32}P - ^{137}Cs) or two radioisotopes of the same element like (^{22}Na - ^{24}Na), (^{55}Fe- ^{59}Fe), (^{89}Sr - ^{90}Sr), (^{90}Y - ^{91}Y), (^{110}Ag - ^{111}Ag) are used.

Advantages of Double Labelling

The simultaneous use of two radioisotopes in an experiment is made feasible by the fact that measurements of each in the same sample are possible. In general, double labelling offers a means of comparing directly in a biological system the movement, synthesis, or degradation of two substances or the differential behavior of two parts of a given molecule. Double-isotope studies, by allowing two sets of data to be obtained from the same organism, may save considerable time and expense, especially where the experiment involves systems that are difficult to prepare. This is the situation for animals or plants brought to special nutritional status or raised under special environmental conditions.

The synthesis of nucleic acids has been studied by the simultaneous injection of carbon-labelled glycine and P^{32} into rats. The specific activities of the P^{32} and C^{14} in the purine nucleotides from DNA (deoxyribosenucleic acid) were similar, indicating that both precursors were incorporated into some intermediate at the same time. However, the specific activity of the P^{32} in the RNA (ribose nucleic acid) purine nucleotides was considerably higher than that of C^{14}, indicating that the P^{32} was incorporated earlier in this case.

A very powerful method is available in those instances where it is possible to use two different radioisotopes of the same element (Na^{22} - Na^{24}, Fe^{55} - Fe^{59}, Sr^{89} - Sr^{90}, Y^{90} - Y^{91}, Ag^{110} - Ag^{111}). A considerable advantage lies in the correction, for secondary losses, which is an extension of isotope dilution. For example, assume that radioisotope is administered to the system and that an unknown amount of it becomes located in the sample. A known amount of radioisotope B is added to the sample, which is then analyzed for both isotopes. It is clear that the following is true:

$$C = A \frac{Ba}{Br}$$

where C = amount of radioisotope A in sample.

A = amount of radioisotope A measured in sample or in given portion of sample.

Ba = amount of radioisotope B added to sample.

Br = amount of radioisotope B measured in same portion of sample in which A is measured.

In general, the procedure is as follows:

(a) The unknown acids are converted to labelled derivatives by reaction with I^{131}-pipsyl chloride.

(b) A pure sample of each amino acid to be measured is labelled with S^{35} -pipsyl chloride, and a known amount of each derivative is added to the unknown mixture.

(c) The mixture is separated into components on a paper chromatogram

(d) The bands of the paper chromatogram are sectioned, the amino acids eluted from the sections, and the S^{35}/I^{131} ratios determined. The identification of the bands is determined from previous experience with the movement of known amino acids. If a given band is pure; this will be shown by a constant S^{35}/I^{131} ratio in the various sections of the band. Further more the original content of the I^{131} derivative of a particular amino acid can be quantitatively estimated from the S^{35}/I^{131} ratio by the same type of calculation as described above for the iron experiments.

❑❑❑

Chapter 7

Dosimetry

7.1 Measurement of Radiation Dose

Radiation dosimetry involves the measurement of the energy actually absorbed in a substrate exposed to ionizing radiation. The radiation may be either electromagnetic or corpuscular in nature. The passage of electromagnetic radiation through a substrate causes ionization by virtue of the secondary electrons released. Particle radiations on the other hand, ionize directly because they are charged, or indirectly through the charged particle resulting from a collision with a substrate. Alpha, beta particles, and protons yield primary ionization. Gamma rays and neutrons produce ionizing radiation by their interaction with the substrate in passage through it.

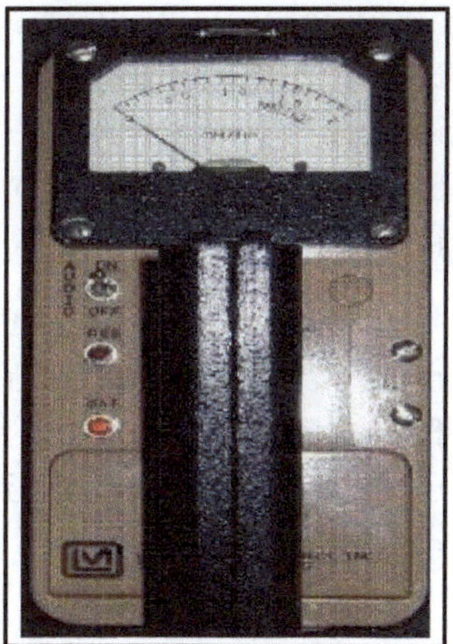

Fig. 7.1. Dose rate meter

Measurement of radiation dose involves the intensity of the source (characteristic solely of the source), the cumulative effect on the substrate and the rate at which the effect is brought about. The source is characterized by the nature and energy distribution of the radiations and the rate at which the radiation is being emitted.

(1 curie = 3.7×10^{10} disintegrations per second).

7.2 Units of Radiation Dose

The original unit of measurement of radiation was the roentgen, r, and was defined in term of ionization events. A roentgen (r) of radiation has been defined as

(1) The quantity of radiation which produces one esu (electrostatic unit) of positive or negative electricity per cubic centimeter of air at standard temperature and pressure or.

(2) The quantity of radiation that will produce 2.083 x 10^9 ion pairs per cubic centimeter of dry air or.

(3) The radiation received in one hour from 1 gram source of radiation at a distance of one meter.

A signal charged ion carries 4.8 x 10^{10} electrical units and one r forms, in one cubic centimeter of air, 2.08 x 10^9 ions of either sign, or 2.08 x 10^9 ion pairs. At standard conditions one cubic centimeter of air weights 0.00129 gram so that one gram of air absorbing one r would result in the formation of 2.08 x 10^9 divided by 0.00129 or 1.61 x 10^{10} ion pairs. The energy required to produce one ion pair in air is known to be about 32.5 electron volts. The energy required to form 1.6 x 10^{12} ion pairs is 5.23 x 10^{13} electron volts.

The gram roentgen is the amount of radiation absorbed per gram of dry air. This is a quantity equal to 89 = 1.3 ergs per gram of dry air. Because the radiation absorption characteristics of moist tissue are different from that of air, the energy absorbed by the tissue would be greater.

The roentgen equivalent physical (rep) was developed to describe the dose rate for primary ionizing radiations, e.g. high energy electrons. The rep is defined as that dose of ionizing radiation, which produces energy absorption of 84 ergs per cubic centimeter of tissue. This value was later adjusted to 93 ergs per cubic centimeter.

The roentgen equivalent mammal (rem) is defined as that dose absorbed by a mammal when exposed to ionizing radiation biologically equivalent to the dose of one r of gamma radiations. The dose in rem is equal to the dose in rep multiplied by the relative biological effectiveness (RBE). The rem is of little value unless the RBE is specified and applied.

7.3 Radiation Absorption does

The rad is the unit currently in widespread use which defines the absorption of energy by a substrate directly. The rad is defined as the absorption of 100 ergs per gram of material. This yields a direct means of measurement which has been found most useful in radiation technology. The rad is approximately 1.1 times larger than the rep found in the literature. The current trend in radiation technology favors the use of the rad unit.

Curie

The curie (C) has been defined as 3.7×10^{10} disintegrations per second (dps) for any isotope (internal congress on radiological units, 7th International congress on Radiology 1953). The amount of carbon-14 required to bring this level of activity may be calculated as follows:

η = Number of atoms/curie

λ = Radioactive decay constant

T = half life

t = mean life = 1/ ë = 1.44 T.

then, N = 3.70×10^{10} seconds or

η = T 3.70×10^{10} seconds.

Taking the half life of carbon-14 as 5600 years (or 1.77×10^{11} seconds)

$$\eta = (1.44)(1.77 \times 10^{11})(3.7 \times 10^{10})$$

$$= 9.45 \times 10^{21} \text{ atoms } C^{14}.$$

But,

$$\eta = N \frac{M}{W}$$

where, M = number of grams/curie

N = Avogadro's number

W = atomic weight

Consequently,

$$M = \eta \frac{W}{N} = \frac{(9.45 \times 10^{21})\ (14)}{6.02 \times 10^{23}} = 0.223 \text{ g}.$$

Intensity of radiation: The intensity of radiation is the energy flowering per unit area perpendicular to the beam per unit time. Units: ergs/Sq. cm/sec. or walts/sq. cm.

Quantity of Radiation

The quantity of radiation is the time integral of the intensity of radiation.

Absorbed Dose

The absorbed dose is the amount of energy imparted to matter at a site by ionizing particles per unit mass of material. The unit dose is a rad =100 ergs/g. However, the dose is usually determined by ionization, rather than calorimetrically, according to the equation.

Em	= SWJm
where, Em	= Energy imparted to unit mass
W	= average energy expended by ionizing particle/ion pair formed in the gas.
Jm	= ionization/unit mass of gas.
S	= ratio of stopping power of material to gas.

The best values of W and S are to be determined experimentally. The integral absorbed dose is then the integral of the energy absorbed in a given volume unit : 1 gram-rad = 100 ergs. to the source. The roentgen does not involve time, so that dosage rates are given in r/t. The maximal daily human γ radiation dose is 100 mr (through source vary as to whether this is over or 8 hour or a 24 hour day) of this amount, cosmic radiation at sea level produces at least 0.2 mr.

7.4 Dosimeters

To be functional, a dosimeter should be independent of the dose rate over a wide range and be relatively insensitive to small temperature and pressure changes. There should be no dependence upon the concentration of materials in the substrate, as they may change during irradiation. Further more, the quality of the radiation applied should not be a determining factor in the response of the dosimeter. It would be desirable to have simple procedures involving no special partification of needed roentgens, and solutions used in dosimeter should be usable in equilibrium with natural surroundings. Chemical dosimeters most often fail in the requirement of independence of the quality of radiation supplied. Non- chemical dosimeters often fail in the quality dose rate, temperature and pressure response, as well as the quality of radiation. Dosimeters have therefore evolved having unique features of special application. Unfortunately functional, simple durable dosimeters capable of evaluating radiations varying in intensity a million fold are not available. However useful dosimeters are available far various dose ranges and will be described below:

Table 7.1: Units of radioactivity measurement

Unit	Definition	Relationship to other units
Radiation absorbed dose (Rad)	100 egs/g (energy imparted to a unit mass of matter by ionizing radiation)	0.87 rad = r.
Roentgen (r)	An exposure dose of X or Y radiation such that 1.61 x 10 ^{12} ion pairs are produced per gram of air	r/0.87= rad.
Curie (C_i)	A quantity of ratio nuclide disintegrating at a rate of 3.7 x 10^{10} atoms per second.	C = 10^3 mc,C = 10^6 μc.
Millicurie (m.c.)	One thousand of curie.	mc = 10 $^{-3}$ C ,mc = 10^3 μC.
Microcurie (μC)	One millionth of a curie :3.7 x 10^4 dis intigrations per second (2.2 x 10^6) disintigration per minute	μC = 10^{-6}C,μ C = 10^{-3}μ C.
Disintegration per minute (dpm)	Number of atoms disintegrating per minute	2.2 x 10^{-6} dpm = 1 μC.
Counts per minute (cpm)	Number of β particles detected per minute by a detecting device	cpm/dpmx efficiency of counting deice.

The establishment of primary standards for the measurement of radiation dose is complex. However, once primary standards are established, secondary references can be devised which usually are more convenient than primary standards.

Dosimeters of three types are actually required:

(1) Primary standards,

(2) Operating standards and

(3) Production control devices.

An example of a primary standard dosimeter is the calorimeter developed at the National Bureau of Standards. Usually primary standard dosimeters do not land themselves to routine usage, and operating dosimeter systems have been developed which have been standardized against primary dosimeters. Modified ionization chambers can be used to monitor a radiation source continuously.

The operating standard dosimeters found useful in routine radiation technology are cobalt glass, ferrous sulfate and ceric sulfate dosimeters. The ceric sulfate dosimeters are functional for doses between 200,000 and 5,000,000 rads; the cobalt glass dosimeters are operative between 200,000 and 10,000,000 rads; and the ferrous sulfate dosimeters are functional between 50,000 rads.

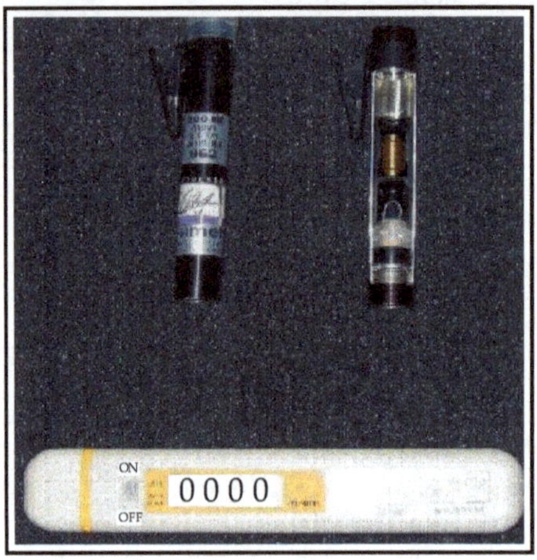

Fig.7.2: Personal Dosimeters

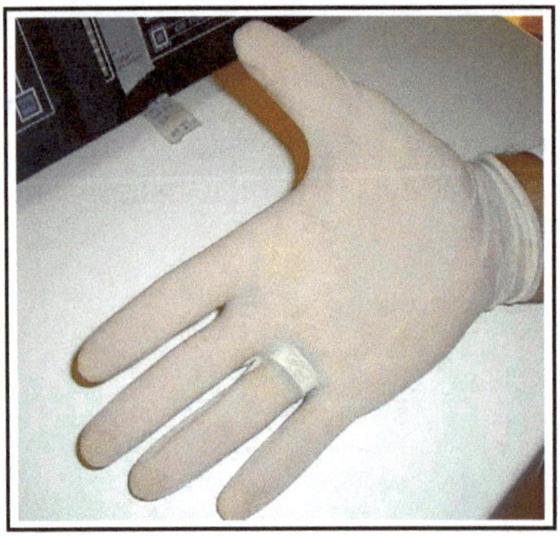

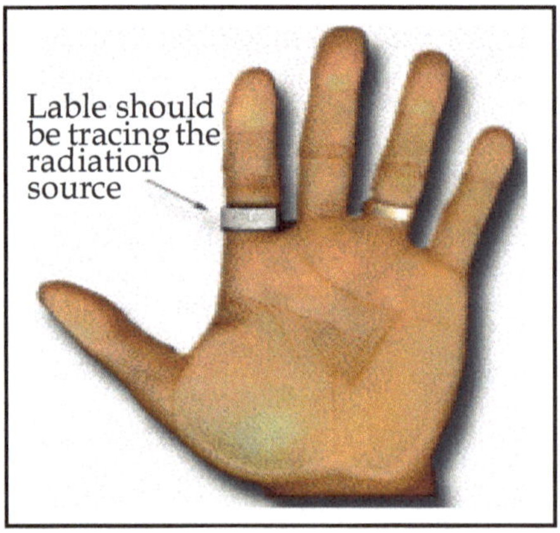

Fig.7.3: Finger Ring Dosimeter

7.4.1 Ferrous sulphate Dosimeters

One of the most widely used standards is the Frick dosimeter, which consist of an acid solution of a ferrous salt containing some chloride to inhibit side reactions and impurity effects.

The formation of ferric ion is the key effect which permits the evaluation of the radiation doses; the concentration of the ferric ion is determined spectrophotometrically by its absorption band at the wavelength of 305 meter. The concentration depend on light absorption is exponential (Beer's law) so that the logarithm of the light automation is proportional to the concentration. On the basis of careful experimentation, it has been found that there are 15.45 of ferric ions formed for every 100 electron volts of absorbed energy, so that the determination of ferric concentration gives a direct measure of the energy absorbed. The yield of ions per 100 electrons volts is called the G-value, and in this case G equals 15.45.

7.4.2 Ceric sulphate Dosimeter

The ceric sulphate dosimeter system is influenced by the same general and specific factors which influence, the Fricke dosimeter. However, the ceric system is not influenced by the oxygen content nor by dose rate effects. The ceric ion concentration may be determined effect by titration with ferrous sulphate or by direct manuscript spectrophotometer of absorption of ceric ion at 310 meter. The latter method is generally used.

7.4.3 Cobalt Glass Dosimeter

High land dosimetry of gamma radiation has been found functional by the use of cobalt glass dosimeters. A silicate glass containing 0.1% CO_3O_4 is manufactured by Bausch and Lamb optional Co., New York, and is available for the manufactures. The chief advantage in using the cobalt glass system over other means of dosimetry are the small size, ruggedners, and case of mounting for mapping dose distributions, and the possible reuse by heating to eradicate the color decreased during the radiation. Because glass tends to fade during storage, the system must be calibrated before use and optical values compared with some absolute primary system of dosimetry such as the ion chamber, the colorimeter, or even a ceric sulphate system. The usable dose range is mealy considered above 200,000 rads and below 10,000,000 rads. Glass dosimeter is capable of high precisions in the release of two per cent. Heating such glasses in muffle furnace for six hours at 450^0 C permits re-use of the glass dosimeter. The use of sandwiches of glass give layers thick is useful establishing electric equilibrium in the system thereby making it dependent of the surrounding material. The assumption is made that the measurements made with essentially one MeV photon which cause secondary electrons to be emitted which will have range of less than 1.5 mm. in the glass.

7.4.4 Go-No-Go Dosimeter

With the establishment of the dose, rate of a radiation source and the dose absorbed in a substrate, it is possible to have available workable monitoring devices which are called Go-No-Go dosimeters. Each unit container of substrate being irradiated should carry such a device. These production control devices may be as simple as colour making on containers or coloured plastic tapes. Here again colour changes are related to dose received. Although the Go-No-Go dosimeters are not as precise as primary secondary standards, there inerpersive controls are extremely useful in positively indicating that radiation treatment was given to the unit package of the substrate.

□□□

Chapter 8

Detection and Measurement of Radioactivity

Although some forms of electromagnetic energy, such as light and heat, can be detected by the human senses. One of the greatest draw backs to high energy radiation is the inability to detect it. We cannot see, feel, taste, smell, or hear the various forms of ionizing radiation. Fortunately, ionizing radiation interacts with matter which makes detection and measurement possible by utilizing specialized equipment. In this section we want to introduce you to the various ways and means of detecting and measuring ionizing radiation.

As mentioned previously, Becquerel discovered radioactivity because it left marks on photographic film as a means of detecting radiation. However, there are more definitive means commonly used by scientists and technicians who study and work with radiation. The equipment utilized for the detection and measurement of radiation commonly employs some type of a substance or material that responds to radiation. Many common methods use either an ionization process or molecular excitation process as a basis. Remember that we stated earlier that radiation interacts with matter. For detection and measurement purposes the process of ionization is the most commonly employed technique, based on the principle of charged particles producing ion pairs by direct interaction. These charged particles may collide with electrons, which remove them from their parent atoms, or transfer energy to an electron by interaction of electric fields.

How do you choose a detection device?

Important considerations for choosing a particular type of detection device include the application, the type of radiation, the energy of the radiation, and the level of sensitivity needed. Remember from previous discussion that radiation exists as waveforms with varying energies and may be either particulate or electromagnetic in nature.

The unique advantage of an isotope is that one can use an extremely small, or trace, amount to follow the behavior of much greater amounts of the abundant isotope of the same element. From this simple basic conception a number of more elaborate treatments may be developed. In considering the principles of application it is necessary to assume that the tracer isotope will behave in the system similarly to its more abundant counterpart, and that the isotope can be conveniently measured either by the radiation that it emits, or by determining its abundance in the case of "heavy" or stable isotopes. Thus we may distinguish the tracer from the common isotope and obtained an estimate of the amount of tracer present. There are situations where these assumptions do not hold, but in general they are valid and serve as the basis for the present chapter.

The special advantages of radioisotopes are that it is possible to differentiate between an element or substance that is already present and that which is added to the system; the sensitivity of the method is high.

Sensitivity of various analytical techniques:

Analytical technique		Limit of sensitivity
Spectrophotometric		10^{15} molecules
Radiochemical	^{14}C	10^{11} atom.
	^{3}H	10^{9} atom
	^{32}p	10^{7} atom.

8.1 Quantitative Approach

8.1.1 Soils and Fertilizer

A. Isotopic dilution method

This is one of the important techniques that are widely used wherein a small amount of the labelled test substance is incorporated in the test material and, after the completion of the test, the test substance is isolated and the isotopic content determined. The amount of dilution that has occurred is a function of the amount of the test substances in the original material. The technique has been used to investigate the contribution of the nutrients from each one of the two sources (say fertilizer and soil) to the plant. In this method, the dilution technique involves the addition of the isotopically labelled substances to the analytical sample. However, if the unknown in the sample could be produced in the labelled form, the procedure can be carried, out according to the same principle and method described earlier.

Detailed classification of radiotracer techniques is given in flow chart as under

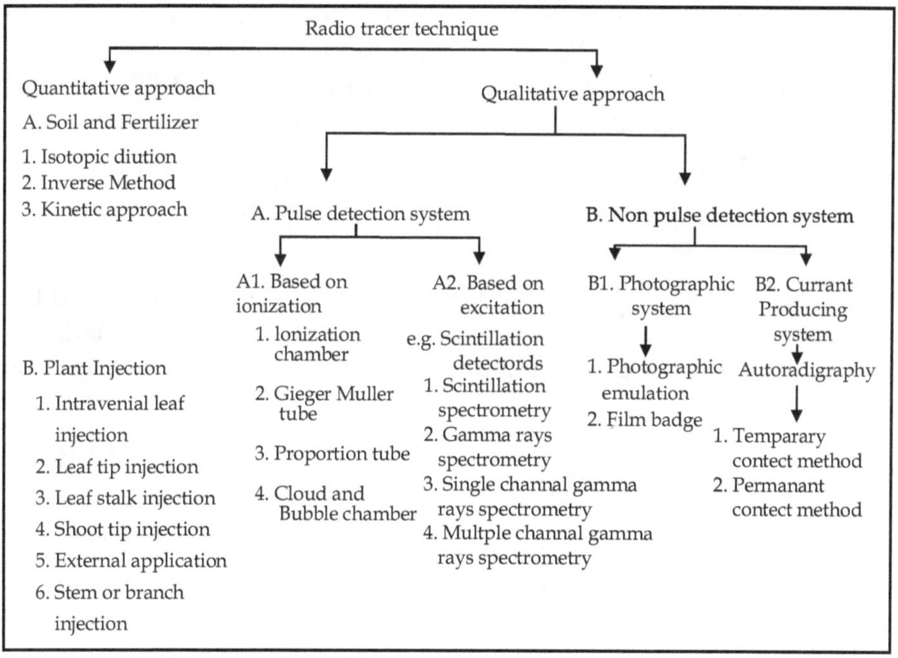

B. Inverse method

This method has certain advantages over the direct method in that smaller amounts of the unknown can be determined and also it is not necessary to have the labelled diluting substance.

C. Kinetic approach

It permits the estimation of the rate of transfer of a substance from one phase to another, determination of the amount present in each phase at a given time and the identification of the processes involved in the study.

8.1.2 Plant injections

It is used for studying the distribution and activity of the root systems of plants and to obtain information on the depletion of nutrients and water in each of the different zones. In the plant injection technique, different methods are used for different purposes. The techniques can be divided into:

A. *Intravenial leaf injection* wherein a small incision is made in the leaf or at a chosen point and the radioisotope is drawn thorough a wick.

B. *Leaf tip injection* wherein the tip of a leaf or leaflets is cut off at right angles to the midrib and immediately immersed in the liquid to be injected. The liquid is drawn in through the cut edge of the leaf as a result of transpiration and travels along the veins and vein lets to replace the tissue transpiration losses.

C. *Leaf stalk injection* wherein the leaf is cut off leaving a short length of the stalk which is connected to the tube containing the radioisotope.

D. *Shoot tip injection* wherein the tip of the shoots is cut off and the cut end are kept with the liquid. The cut should be made far away from the growing point to permit the access of the solution to the differential conducting tissues.

E. *External applications* of the leaf through which the rate of translocation to other parts of the plant is measured.

F. *Stem or branch injection:* technique wherein the branches may be injected independently of the root of the tree through suitably placed holes.

In the root pattern studies the radioisotope is injected into the soil at the specific location relative to the plant, and the amount taken up by the plant at different times and at different depths is determined by analysis of the leaf or root tissues.

In the above studies, information on

(1) The time of arrival of the roots at the given location in the soil which is a measure of the overall growth rate at which the root expands through the soil column,

(2) The seasonal and soil effects on the root pattern,

(3) The time and manner of plant competition that may lead to spacing recommendations

(4) Effects of the depth of tillage

(5) Relative importance of various root segments for nutrient uptake,

(6) The efficiency of the root system of a given species in utilizing added fertilizer under given conditions can be obtained.

In all the above techniques, when quantitative aspects have to be studied, the specific activity that is the amount of radioactivity per unit weight of the element present (which includes both active and stable isotopes) has to be calculated.

8.2 Qualitative Approach

8.2.1 Ionization and Excitation

Radiation, whether ionizing radiation or visible light, acts on the electronic structure of atoms. If the radiation, either particular or electromagnetic, is of sufficient energy, it may cause the entire removal of an orbital electron from the atom. This is called ionization. The subsequent history of the detached electron and of the parent atom may assume one of several forms and will be discussed below.

If the energy of the radiation absorbed by an atom is not sufficient to result in ionization of an electron, it may result in a perturbation of the otherwise stable interrelations between the nucleus and the orbital electrons. This phenomenon is called excitation. If the electron continues to absorb energy from repeated exposure to radiation, it may be cited sufficiently to break the bonds of stability holding it in balance and to jump free of the atom. This separation of an electron from the atom marks the boundary between excitation and ionization, both of which are essentially of the same nature. The two conditions differ only in that the accumulated energy necessary to result in ionization is greater than that resulting in excitation alone.

Excitation of an atom, under the influence of radiation, commonly manifests it self in the emission of electromagnetic radiation during which the excited atom seeks a level of stability. An alternate method of loss of excess energy is by direct transfer of excitation energy to an adjacent atom.

Either excitation or ionization may be induced in an atom by particulate radiation, commonly by alpha or beta particles or by electromagnetic radiation; either particulate or electromagnetic radiation capable of inducing ionization is designated as ionizing radiation. For electromagnetic radiation, the limits of ionizing radiation are within the spectrum of ultraviolet light. Energetic ultraviolet light, with energies of 7 to 12 ev, corresponds to wave lengths between 1700 and 100 A and is capable of producing ionization: these wavelengths are immediately longer than those of X-rays (which also produce ionization) and shorter than ultraviolet radiation near the visible spectrum.

8.2.2 Nuclear Radiation Detectors

The quantitative determination of radioactivity depends on the ionization or excitation of matter by the radiations emitted by radioactive elements or by other sources of electromagnetic radiation. It is apparent that the usefulness of any radioisotope as a tracer lies in the detection, recording, and measurement to provide a system for such quantitative determination. Although significant advances in the development of instruments have been made not all are applicable to the study of bone and teeth. Selection of the method to be used depends upon technical requirements as well as upon financial considerations. The scope of this volume permits only a brief description of the basic principals of the systems limited to tracer techniques and excludes those used in radiation therapy and dosimetry. Adequate treatment of instrumentation, including characteristics of counting, can be obtained from references on methodology.

For practical purposes, the nuclear-radiation detectors may be divided into two general categories:

(1) Pulse systems, in which the output of the detector represents individual interactions of a nuclear particle or a gamma ray with the detector; and

(2) Non-pulse detection systems in which no attempt is made to resolve the individual interactions of radiation, so that the output represents the average or total cumulative effects.

8.2.2.1 Pulse systems

Pulse systems may be further divided into:

(I) Those using the principle of ionization and ion collection, such as ionization chambers, proportional and Geiger-Muller tube counters and

Based on principle of ionization or ion collection electroscope - use of ionization principle

All ion-collection detectors have the basic features of an electrode system consisting of a circuit through which the current passes when ions, collected in a gas medium, are attracted to the charged electrodes. The cathode and anode are connected through a resistance which permits a potential difference to be applied between them. The resistance may be inserted into the circuit or it may be the parts of the circuit.

When a charged particle passes through the gas, electrostatic interaction occurs between the particle and the orbital electrons of the gas molecule. Ion pair production results from the splitting of the molecules of gas along the path of the particle. The number of ion pairs formed per unit path length is the specific ionization and depends on the charge and speed of the particle. The electrical charge collected after passage of the particles varies owing to the different numbers of ions formed by them as they traverse the gas in the collecting system. For example alpha particles, being doubly charged and relatively slow moving, will produce a specific ionization greater than that of beta particles. Gamma rays and X-rays do into ionize gases directly since they are not themselves charged. However, ionization occurs by the secondary beta particles produced by their absorption, either in the gas itself or in the walls of the counting vessel.

The difference in types of instruments depend primarily on the magnitude of the voltage applied to their electrodes and are best described in terms of the region of operation, as illustrated in Fig.8.1.

Ionization Chamber Region

In region II, ion chambers can be operated as pulse systems by counting individual events. Ionization chambers can also be operated as non-pulses detection systems, i.e. as current devices. These devices will be described in a later section.

At low voltage a very small fraction of ions formed in the detector can be collected but with increase in potential the field will increase proportionately so that all the ions produced inside the detector are collected. With increasing voltage, more ions are collected before they are able to recombine; this proportion increases until all ion pairs that are produced can be collected. A further increase in voltage will have no appreciable effect on the collection efficiency or nor the current in the circuit; this voltage range is the saturation voltage or saturation plateau. It is in this region that an ionization chamber is normally operated.

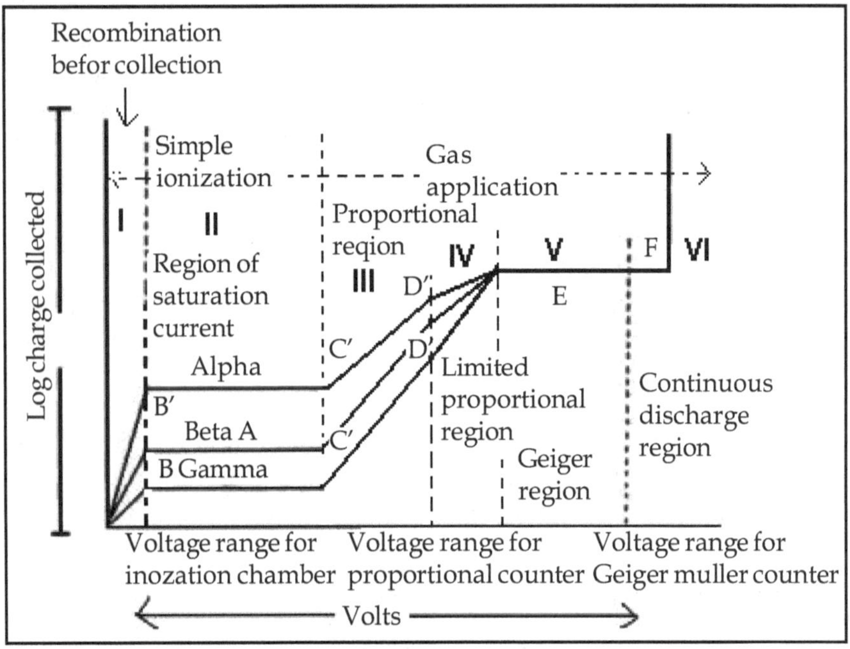

Fig. 8.1: Curves of pulses height v/s applied voltage to illustrate ionization, proportional and Geiger Muller regions of operation.

Proportional Region

When counter tubes are operated in this region (Fig. 8.2), it merely indicates that the size of the electric pulses (output pulse) is proportional to the number of electrons released by the particles. All primary ion pairs produced are being collected at the electrodes. With higher potentials, some of the primary electrons are accelerated in the field of the electrode, to such a point that they interact with other gas molecules, thereby producing secondary ionization. Gas amplification occurs when more ions reach the collecting electrode than were in the original collisions and the charge collected increases over the saturation value. The increase in the charge collected at the electrode is proportional to the voltage. The number of ions collected will remain proportionately different for alpha and beta particles, owing to the different number of primary ions formed by the radiation. The discharge is localized to a small portion of the tube in this region.

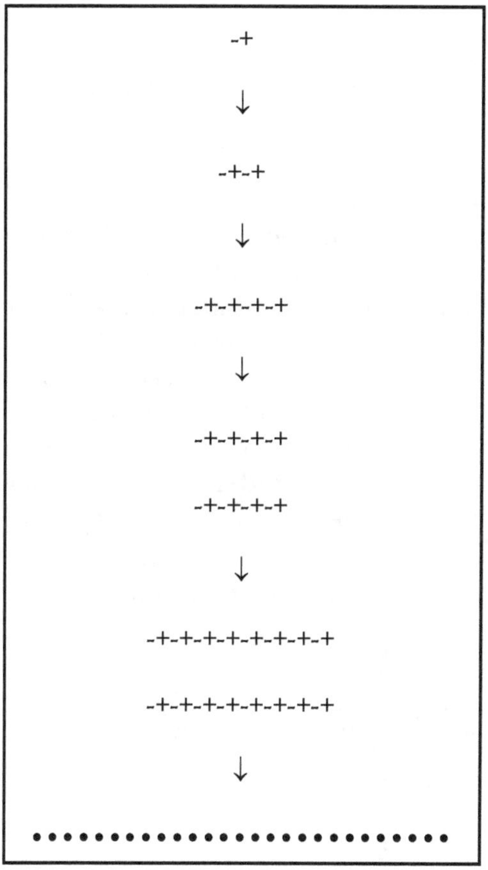

Fig. 8.2: Proportional region

Geiger Region

In region V the charge collected is independent of the alpha and beta radiation and the number of primary ions initially formed all ionization, regardless of its nature, will produce similar end results. The detector used in this high-voltage region is the Geiger-Muller tube counter. The tube itself is a gas-filled (helium or argon) cylindrical cathode and an axial wire anode. A high electric voltage, approximately 1000 volts, is applied between the electrodes. The field intensity about the axial electrode is so high that any ion formed, whether of primary or secondary origin, can be accelerated enough to cause additional ionization of the gas molecules. A chain reaction is created, and an avalanche of electrons and positive ions will form unless the tube is quenched. Most tubes used in biology are self quenching; i.e., they are filled with a mixture of a noble gas and an organic vapor. The discharge is stopped after a fraction of a second, and the tube is then prepared to respond to another impulse; this interval is the dead time and, for a typical Geiger-Muller tube, is of the order of 200 to 300 μsec.

A. Ionization chamber

When ionization occurs in air, the air becomes drastically conductive. Roentgen investigated this effect in his early experiments on x ray by using an apparatus similar to that shown in Fig.8.3 The radiation enters the window and causes ionization within the chamber. The electrons drift towards the positive electrode and the positive ions towards the metal container. This is occurs a current in the circuit that registers on a sensitive a-meter. The ionization chamber gives a measure of the strength of the radiation entering its air space, but it does not record the passage of each particle individually.

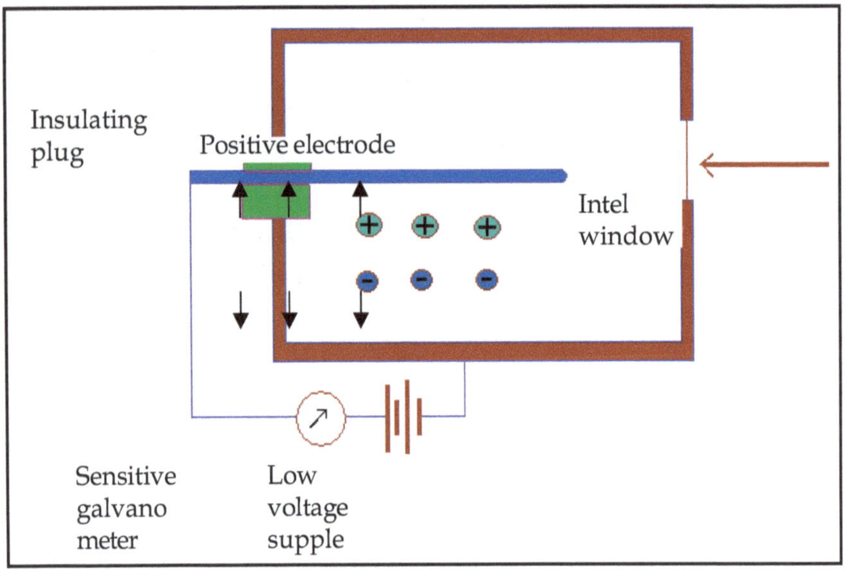

Fig. 8.3: Schematic diagram of an ionization chamber.

B. Geiger-Muller Counter

The Geiger-Muller Counter evolved from the ionization chamber. Instead of air at atmospheric pressure, the Geiger tube (Shown in Fig. 8.4) contains argon at a pressure of about 0.1 atm. Also, the potential difference between central electrode and the walls of the tube is about 1KV instead of the relative low voltage used in the ionization chamber. These changes amplify the ionization process. Ionizing radiation enters the tube and ionizes the low-pressure argon along its path. Because of the high voltage between the electrodes and the low gas pressure, the argon ions achieve sufficient energy to cause additional ionization as they rush towards the electrodes. This type of gas discharge is called an avalanche. The current pulse registered by the avalanche discharge, which is much larger than that due only to the ions formed directly by the radiation, causes a voltage pulses across the series register, which activates the display unit and records the passage of the ionizing particle. The end window of a Geiger-Muller counter would have to be exceedingly thin to admit α particle, but β particles are readily admitted and counted. Although γ rays can enter both the end window and the side walls, they do not cause much ionization in the gas. Their detection is usually indirect. They reject photo-electrons from the inside a wall of the tube which causes an avalanche that is countered in the usual way.

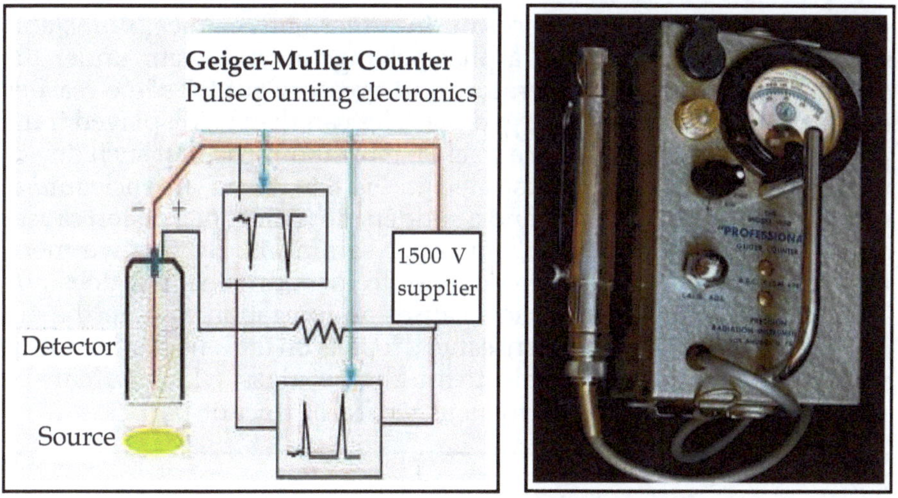

Fig. 8.4: Schematic diagram of a Geiger Muller counter.

The Geiger-Muller counter is rugged and portable. It is especially suitable for detecting β radiation, however, it can not count the β rays if they come at too fast a rate, because it takes about a millisecond for an avalanche to form, register its current pulse, and clear the ions from the tube. If a second ionizing particle arrives during the avalanche caused by a previous particle, it simply adds to be avalanche and does not cause a clearly identifiable discharge of its own. Consequently, the two are not counted separately. This effect limits the Geiger- Muller counter to about 1000 counts per second.

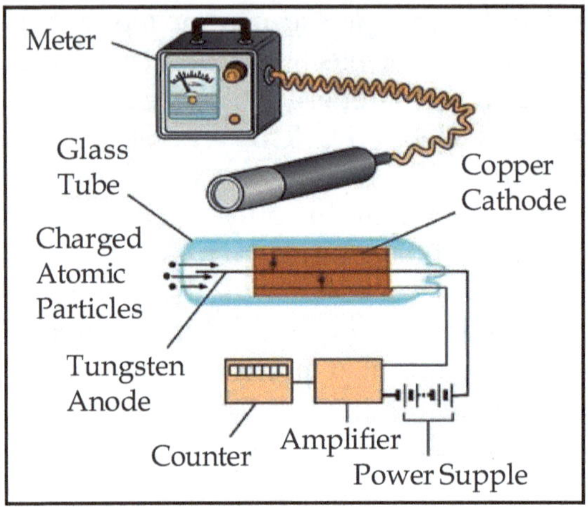

Fig. 8.5: Geiger-Muller counter

C. Cloud and Bubble Chambers

Certain systems can be set up in a thermodynamically unstable state. By this we mean that a phase change should occur under the existing conditions of temperature and pressure but for same reasons does not. The passage of charged particles can then be displayed if the phase change occurs along the track of ionization. The first such device was the cloud chamber, illustrated in Fig.8.6. Air in an enclosure is saturated with water vapor, and a sudden movement of a piston causes a pressure change that result in super saturation. By this we mean that the water vapor should condense to form droplets but does not. The passage of charged particle causes an ionization tract in the air, and the water begins to condense into droplets on the ions. The resulting vapor trail, which is visible through the glass window, can be photographed to form a permanent record of tract of particle.

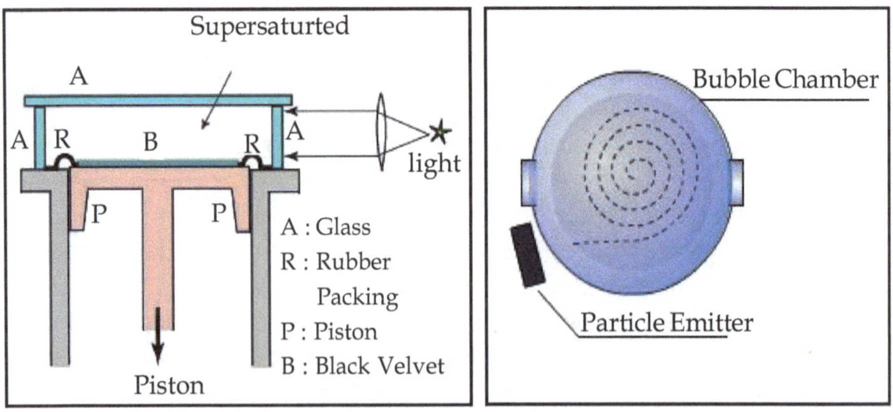

Fig. 8.6: Schematic diagram of a cloud chamber detector

Bubble chambers operate in similar fashion Liquid hydrogen is superheated to the point where it should vaporize but does not. The ionization tract of a particle causes the vaporization to begin, and the trait is visible as tiny bubbles of gas. A magnetic field is often used to cause curvature in the tract of charged particles.

(II) Those using the excitation principle such as the scintillation detectors

When electromagnetic (gamma rays or X-rays) radiation interacts with matter, ionization is also produced, but it occurs by any of the three possibilities. These are: the photoelectric effect, the Compton interaction (Compton scattering), and pair production. The method by which ionization occurs by any of the possibilities depends on the energy and velocity of the charged electron and the atomic weight of the matter (absorber).

In the photoelectric effect where there is low energy and the absorber is of high atomic weight, all of the energy of a gamma-ray photon is transferred to the ejected electron. In the Compton interaction, only part of the energy of a medium-energy gamma-ray photon is transferred to the electron. This interaction will occur in absorbers of any atomic weight. In pair production, the high-gamma-ray energy goes into the formation of a positron and a negatron and imparts kinetic energy to them; this occurs in absorbers of high atomic weight. These electrons then lose energy by ionization and excitation until they are stopped. The positron then combines with an electron and both particles are totally converted to annihilation quanta or gamma rays of electromagnetic energy.

There are many chemical substances, which absorb excitation energy and re-emit the energy in the form of light photons. Such substance is called phosphorus or scintillates. When radiation interacts with phosphorus, kinetic energy of particles of electromagnetic radiation is converted into flashes of visible or ultraviolet light. The total amount of light emitted is proportional to the energy absorbed by the phosphorus. The light emitted is close to the limit energy to flashes of light is scintillation.

A. Scintillation counter

In his early experiments Rutherford detected α rays with a screen coated with a layer of small zinc sulfide crystals. When struck by an á particles, the zinc sulfide emits a flash of light, which Rutherford observed through a magnifying glass. The same principle is used in a TV screen. When fast electrons stick the layer of fluorescent material, it emits visible light. The modern-day version of the scintillation counter uses a single crystal of some suitable substance to absorb the incident particles for radiation and emit visible light. Sodium iodine is frequently chosen as the crystal. A typical arrangement is shown in Fig.8.7. The emitted light passes to a photomultiplier tube, which is an electronic device that converts light into a pulse of electron current. A sensitive tube can convert one photo into a substantial current pulse. The current pulse then activates the display unit, which could be an electronic counter that records the total number of events or a small speaker that gives an audible signal whenever an event is recorded.

The scintillation counter is a ragged instrument that efficiently detects both γ rays and charged particles. It is also energy sensitive detector, since the magnitude of the output is usually proportional to the energy of the incident radiation.

Scintillation counters may be operated in the region of 600-1300-volts, although the majority of photomultiplier (PM) tubes are customarily operated between 850 and 1000 volts. Unlike, G-M and proportional gas counters, scintillation detectors do not have a clearly defined operational voltage. As has already been noted, the size of the pulse induced by a radiation photon or particle which enters a scintillator is proportional to its dissipated energy. Further, if the high voltage is systematically increased then linear amplification results in the proportional amplification of all pulses. Thus the largest pulses due to the most energetic radiation are detected first, but with increasing voltage and hence increased amplification the smaller pulses reflecting the less energetic, and scattered radiation and photomultiplier noise are also detected.

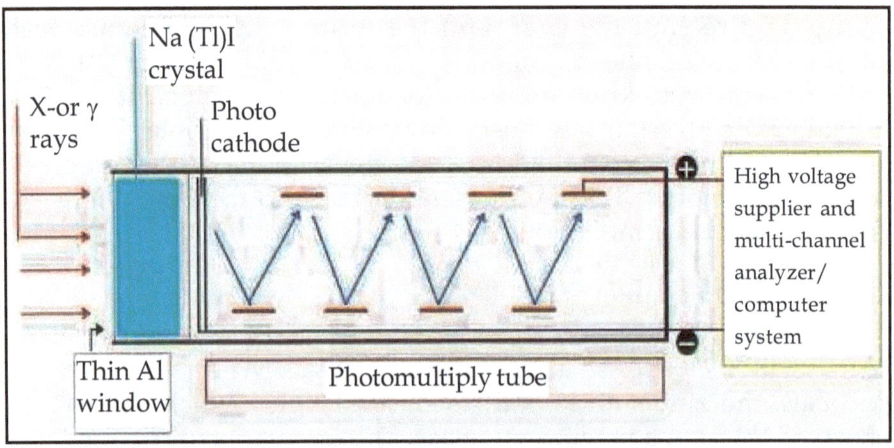

Fig. 8.7: The key components of a typical scintillation counter

In the earliest liquid scintillation counters, a transparent container containing the sample was coupled optically to a single photomultiplier tube. The container was surrounded by reflectors to increase the collection of light by the photocathode. The signal from the photomultiplier tube was directed through a preamplifier, amplifier, and pulse-height analyzer and recorded by a scaler (Fig. 8.10). To reduce electronic noise which masked the signal at room temperature, photomultiplier tube and preamplifier were operated at reduced temperature (0-5^{0}C).

In addition to two photomultiplier tubes and a coincidence circuit, recent liquid scintillation counters contain a pulse-summation circuit. This circuit increases the amplitude of the signals representing radiation interactions by adding the signals from the two photomultiplier tubes. Pulse summation also provides a more narrow and reproducible pulse-height spectrum for the β particles being counted. A block diagram of a modern liquid scintillation counter is shown in Fig. 8.11.

A sample or "cocktail" prepared for liquid scintillation counting consists of at least three components: the radioactive material, an organic solvent and an organic fluor. Additional components may be present to increase the solubility of the radioactive material, to enhance the transfer of energy from solvent to fluor molecules, or to match the fluorescence of the cocktail with the spectral sensitivity of the photomultiplier tubes used to measure the fluorescence. Since the radioactive material is mixed intimately with other components of the cocktail, radiation interactions in extraneous materials are reduced to a minimum: consequently, radiations with a very short range can be detected. For example β particles from ^{3}H, ^{14}C, and ^{35}S can be detected with reasonable efficiency by liquid scintillation counting. The technique of liquid scintillation counting is outlined in Figure 8.12.

These counters have the advantage that the radioisotope is dissolved in the liquid organic scintillator and there is not undetected absorption by the sample of the β-particle emitted. Hence this is used for counting low energy β-emitters like ^{3}H, and ^{14}C which emit β-particles of less than 0.02 and 0.2 Mev., respectively.

Organic scintillators exhibit a very short decay time of fluorescence light produced, and therefore resolving time corrections are not normally necessary. In addition, these counters can be readily adapted to automatic sample changers. In this method the size of output pulse is proportional to the energy dissipated by the β-particle in the scintillating liquid.

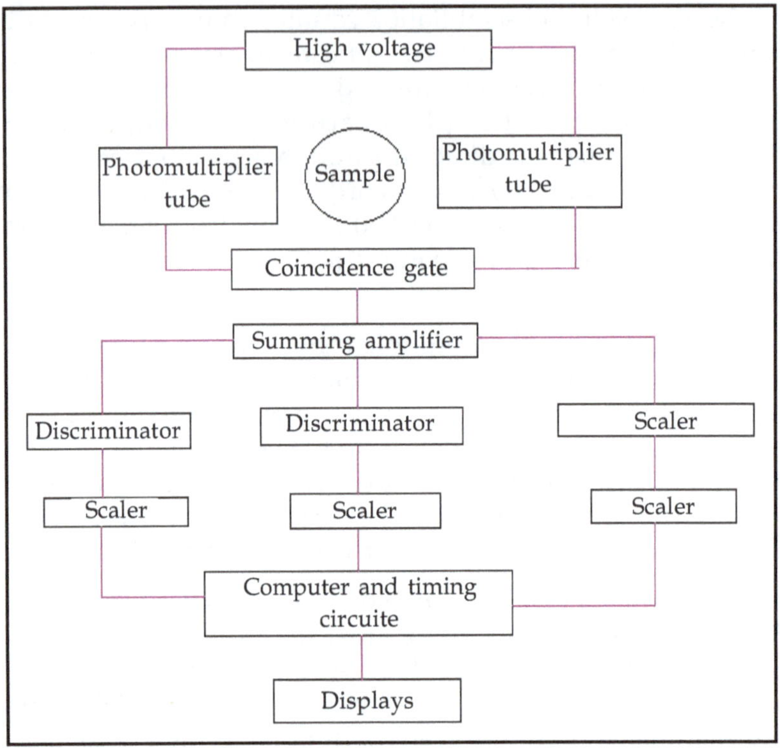

Fig. 8.8: Block diagram of the basic electronic components of a scintillation spectrometer.

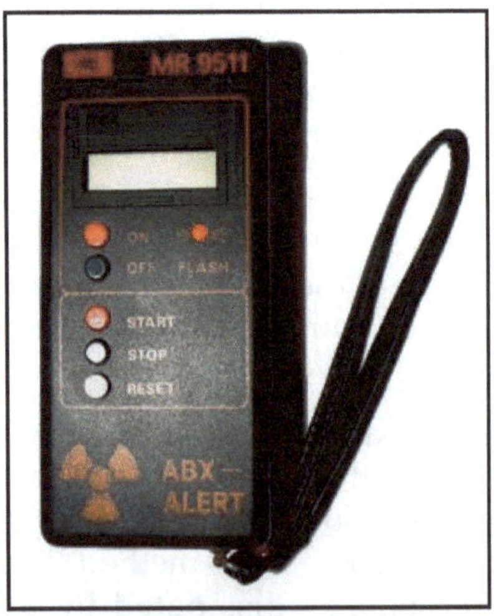

Fig. 8.9: Scintillation counter

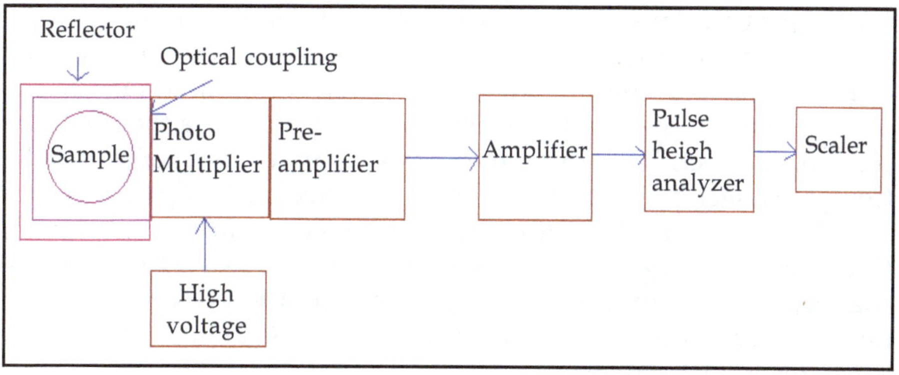

Fig. 8.10: A liquid scintillation counter with a single photomultiplier tube.

B. Gamma-ray spectrometry

The mechanisms of interaction of gamma radiation with the semiconductor detector are similar to those which occur with solid scintillation detectors discussed subsequently. The gamma rays dissipate their energy within the semiconductor depletion zone principally via the photoelectric effect, Compton effect and pair production, with the photon energy appearing as the kinetic energy of the photoelectron, Compton electron and electron positron pair, respectively. The kinetic energy of these electrons is then absorbed by the semiconductor to produce the electron hate pairs necessary for charge collection and counting.

The semiconductor detector offers superior energy resolution to that of the NaI (T_i) solid scintillation crystal detector. This is due in part to the smaller amount of nuclear radiation energy required to produce an ion pair in a silicon (3.6 cv per ion pair) or germanium (2.8 cv per ion pair) semiconductor detectors consequently results in a greater production of ion pairs are also more efficiently collected by semiconductor detectors, which results in the production of signal pulses with magnitudes of smaller statistical variance (greater resolution).

In addition to the greater energy resolution, the semiconductor also produces optimum peak-to-Compton ratios, i.e. the optimum ratio of the full energy peak height of the compton peak height of a gamma-ray spectrum. The full energy peak results when monoenergic gamma rays dissipate all of their energy in the detector material via the photoelectric effect, Compton scattering effects and pair production. The pulses which results from full energy dissipation of the gamma rays appear in the full energy channel of the radiation analyzer. In addition to full energy dissipation, gamma rays may lose only a fraction of their energy in the detector material *via* the Compton scattering effect. Gamma rays may enter the detector material via the Compton scattering effect. For example a deposit part of its energy via the Compton effect, and then escape from the detector. This partial energy dissipation via the Compton effect produces a rather broad peak (because the Compton scatter electrons are not monoenergic) at energy channels inferior to the full energy channel. The peak to Compton ratio of solid scintillation counters is rather low (5:1) in comparison to the semiconductor detector.

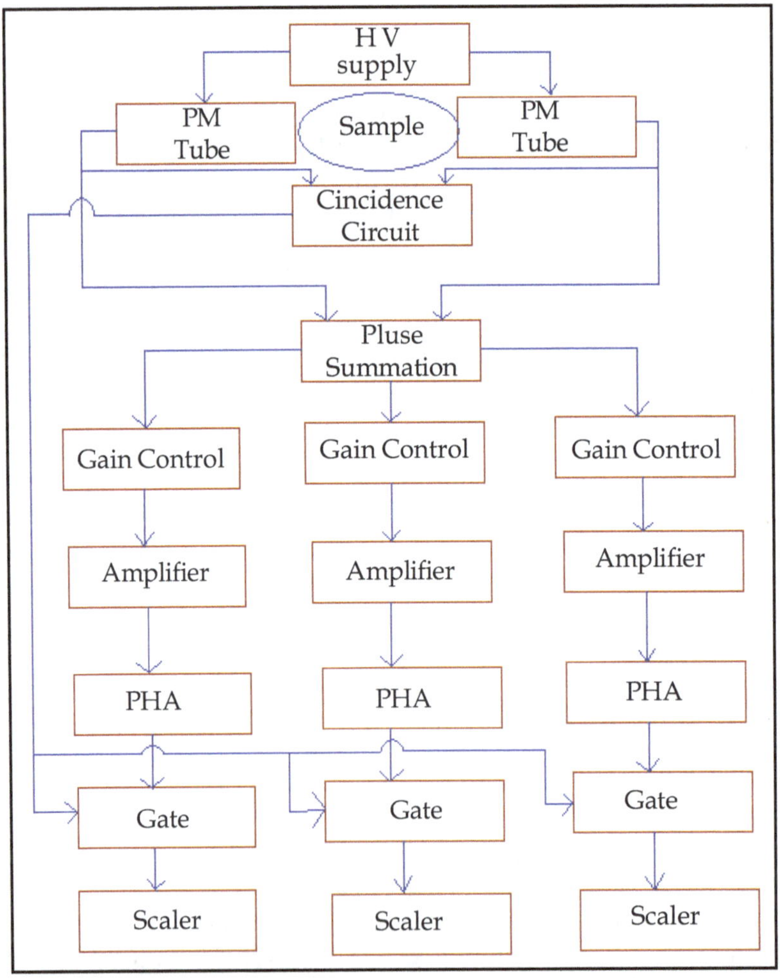

Fig. 8.11: A modern liquid scintillation counter with a coincidence circuit and pulse summation

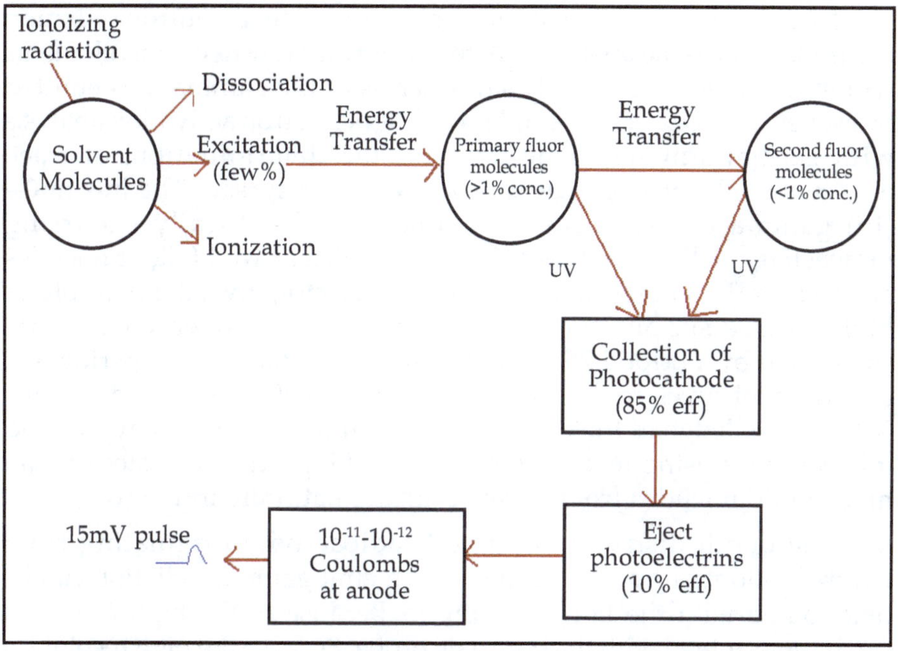

Fig. 8.12: Formation of a voltage pulse in liquid scintillation counting

The efficiency of a Ge (Li) semiconductor detector is conventionally defined as the ratio of the percentage of ^{60}CO gamma rays (133 MeV) which lose all their energy in the semiconductor detector (counts in the 1.33 MeV peak/total number of gamma rays emitted) to the percentage of ^{60}CO gamma rays which lose all the their energy in a 7.62 cm x 7.62 cm (3 in x 3 in) NaI (Ti) solid scintillation detector. The two measurements are made with the some ^{60}CO source (National Bureau of standards) at a distance of 25 cm from the detector. The photo peak efficiency of the Ge (Li) detector is inferior to that of the NaI (Ti) detector because this measurement is a function of the active value of the detector e.g. on efficiency of 7.5%. Consequently, in experimental applications which involve routine sample counting with G NaI (Ti) crystal is preferable. However, in applications involving radiation analysis via gamma ray spectrometry are such as neutron activation analysis. (1) The superior energy resolutions and (2) peak to Compton ratios offered by the semiconductor detector one most advantageous.

The germanium detector has been very useful in the analysis of complex gamma ray spectra encountered in neutron activation analysis.

Neutron activation analysis involves firstly then neutron radiation of the material to the analyzed. After capturing slow neutrons, elements within the material are analyzed in terms of the gamma rays emitted by excited nuclides in the sample. The use of neutron activation analysis with germanium semiconductor for the determination, of trace elements in all sorts of materials is now wider spread. The in-situ Ge (Li) gamma ray spectrometric analysis of naturally occurring radioactive nuclides such as ^{40}K, ^{236}U and ^{232}Th as well fallont nuclides are such as ^{137}Cs in the soil. The Ge (Li) detector, crystal, preamplifier high voltage supply and 7.5 liters of liquid nitrogen coolant are suspended by a pripod 25 cm above the soil surface. The experimental arrangement gives results over a wide soil surface area and volume with a knowledge of the depth distribution of soil radio activities and it is very promising method for distinguishing exposure rates of man made radio nuclides from those occurring naturally in the soil.

The depth distribution of soil radioactivity originating from natural, fallont or tracer nuclides which emit gamma radiation can be determined with the bore hare probe. Because of the high radiation can be determined with the bore hale probe. Because the high resolution Ge (Li) semiconductor detector requires coating at temperatures of 77-90^{0}K to maintain the intrinsic layer width and detector properties constant,. The large Dewar flask required to hence the liquid nitrogen coolant has passed a serious impediment to the use of these detectors in the soil probes. The saturation to this bulky problem has been the use of a canister of prefrozen propane or fluorinated hydrocarbon refrigerant. The detector is coated at the melting point of solid oxygen rather than at the boiling point of liquid oxygen. In practice, the soil probe is lifted from the borchate just prior to the mating of the last trace of solid oxygen. Once back on the soil surface the probe is disconnected at the cryostat and the warmed consister is removed and replaced by a prefrozen conister of cryogen. The probe is then reassembled for further use. The consiters can be changed within 10 min. leaving the probe ready for use with a detector temperature stabilized at 86^{0}K. Schneer et al. (1980), reported that the use of a similar probe for the neutron activation analysis of soil elements and the potential application of the probe in waste products monitoring in the soil.

C. Single channel gamma-ray spectrometry

The positive linear correlation between the gamma ray energy dissipation within the crystal detector and pulse height amplitude output from the PM tube, preamplifier and linear amplifier may be utilized to measure the gamma rays energies of a given radio nuclides. For this purpose a single channel analyzer may be employed.

Pulse Height Analysis

Allows spectrum to be acquired

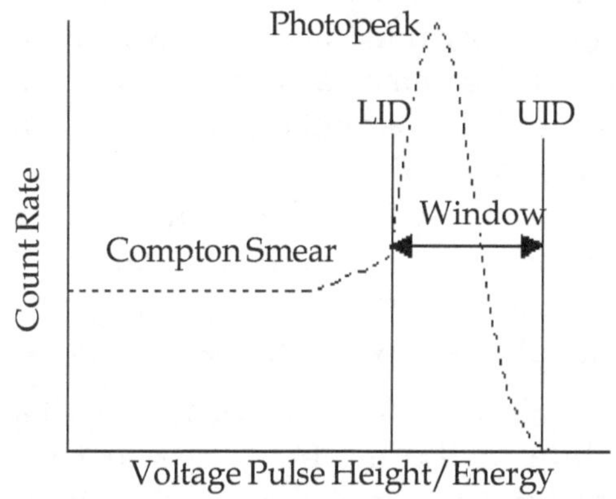

Lower Level Discrimination
Passes voltage pulses > than its setting [ULD]
Upper Level Discrimination
Passes voltage pulses < than its setting [LLD]
=> Variable width **window** which can be placed anywhere along the spectrum or used to scan the spectrum e.g. to detect the photo peak only => ideal gamma-ray spectrum:

A single channel analyzer contains two electronic discriminators set at two variable pulse amplitude levels.

It often referred to as discriminator of lower level and upper level as described previously for liquid scintillation. The pulses which have amplitudes below the level of discriminator lower level are rejected and thus do not produce output pulses only those pulses with amplitudes falling. Within the lower and upper discriminator levels, referred to as the discriminator window, produce output pulses which pasts onto the scalar to be registered as counts. The difference in levels between discriminator two levels (upper and lower in volts) is referred to as the window width, and the level of discriminator lower level is known as the base level of window height.

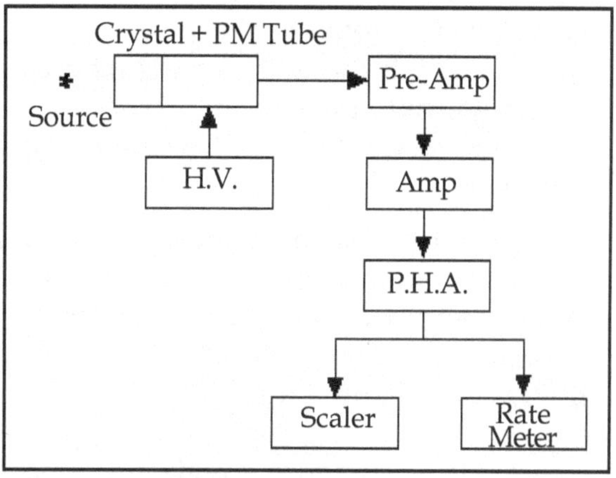

Fig. 8.13: Single Channel Analyzer

The spectrum of pulse heights in volts, being directly proportional to the gamma-ray energy dissipation within the crystal detector, gives the gamma ray energy spectrum of the radiation source. The gamma rays spectrum is simply a plot of relative abundances or population of pulses of a given magnitude (voltage). The portion of the peak enclosed within the 10V window. The distribution of pulse amplitudes found between the lower and upper level discriminators. This peak is the result of the interaction of 0.662 MeV gamma rays of ^{137}Cs with the crystal detector via the photoelectric effect, or a combination of compton effects resulting in the total dissipation of the gamma ray energy in the crystal. Gamma ray spectra are very useful in the identification of unknown gamma ray emitting nuclides. For example, the peak due to the photoelectric effects, usually the highest energy peak, gives a measure of the gamma ray energy of unknown nuclides.

Gamma ray spectrometry may also be applied to assay more than one gamma-ray emitting nuclide in the mixture. It is at times advantageous to employ more than one radiotracer in a given experiment. If these nuclides have distinct gamma-ray energies whose photoelectric peaks are well separated they may be assayed separately by alternately focusing the discriminator window directly over the peak of each nuclide. However, the analysis of many gamma-ray emitting nuclides in a mixture is most effecting carried out with the high resolution obtained with semiconductor detectors employed with multi channel gamma-ray spectrometers.

D. Multi-channel gamma-ray spectrometry

With the single-channel analyzer previously discussed pulses are counted in numerous channel positions corresponding to definite pulse height ranges in order to construct energy spectra. Multi-channel analyzer permits the simultaneous accumulation of pulses in many thousands of energy levels (Channels) in a semiconductor core memory.

The voltage pulses, which results from gamma-ray interactions with the radiation detector, pass on to an analogue to digital converter (ADC) where each pulse amplitude is converted into an equivalent time interval by the production of an equivalent number of pulses from a periodic oscillator. The number of pulses produced by the periodic oscillator is proportional to the amplitude of the converted voltage pulse. An address register then counts the number of pulses produced by the periodic oscillator, which gives a quantified measure of the amplitude of each inputs signal. The quantification identifies the channel among many thousands within which the pulse is to be stored in the core memory. In some analyzers as many as 10^{20} counts may be stored in each channel. The contents of the core memory are viewed on an oscilloscope and if desired, they may be registered by a printer or recorder. Such a simultaneous accumulation and read out of pulses in many thousands of channels corresponding to pulses of defined amplitudes is very time saving, an advantageous which is often necessary for isotopes of very short half life. This advantage offered by multi-channel analyzers can be particularly appreciated when high resolution semiconductor detectors are utilized and several thousands of channels with narrow window widths are required.

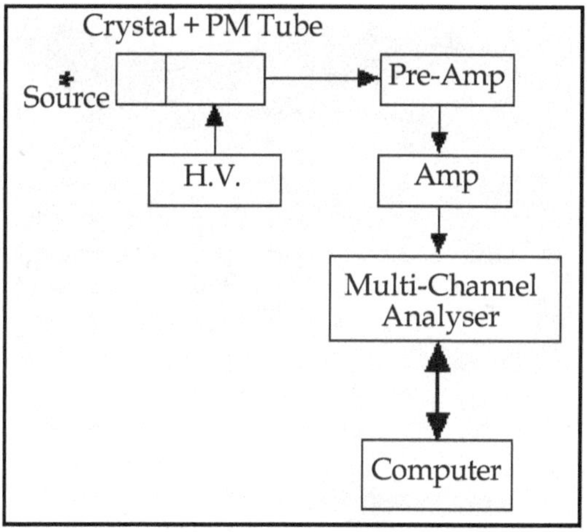

Fig. 8.14: Multi-Channel Analyzer

E. Mass spectrometry

Principle

A mass spectrometer designed for gas analyses essentially is comprised of five units.

(i) An *inlet system* for the introduction of nitrogen gas:

(ii) An *ion source* where the nitrogen molecules, bombarded, with electrons, become charged and accelerated

(iii) A *magnetic field* in which the charged molecules are separated into different paths according to their momentum, the deflection is proportional to the mass of the charged particle as illustrated in Fig.8.15 Where $m_1 < m_2 < m_3$

(iv) A *collector* placed at the end of the fight tube, where the ions discharge and the discharge current are amplified.

(v) A *recorder* which registers the amplified discharge currents as peaks as the mass recorded which register the implied discharge currents as peaks as the mass range is scanned. The heights of the peaks are proportional to the amounts of the three molecules $^{28}N_2$, $^{29}N_2$, and $^{30}N_2$ in the gas mixture. B_7 simply measuring the peak heights, the percentage ^{15}N abundance can be determined.

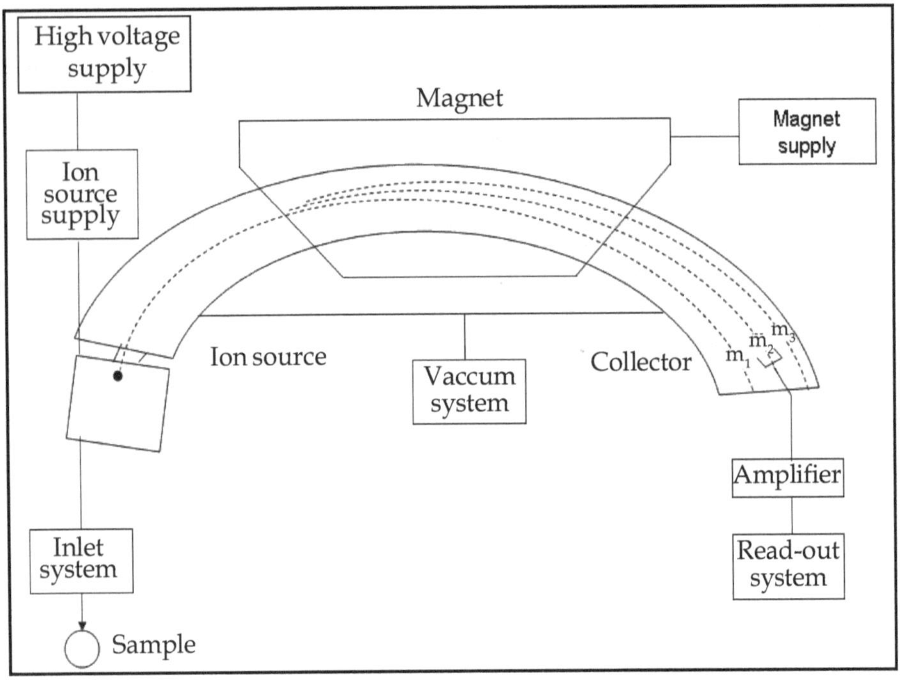

Fig. 8.15: Block diagram of a mass-spectrometer for gas analysis

Although the principle is simple, the practice of mass spectrometry is complicated. The whole system must be maintained under high vacuum, and the electronic components must be highly stabilized. Artifacts can be produced easily when such in instruments are not supervised by highly qualified personnel.

Various types of mass spectrometers are available. The single focusing mass spectrometers depicted in Fig. 8.15 is most commonly used for nitrogen analysis special application require two stage and double focusing mass spectrometers as well as time of flight, cycloidal focusing and radio frequency instruments. The latter are mass filters better known quadrupare spectrometers.

The gas sample is introduced into the gas inlet system, where the pressure is reduced, the sample gas passes into the ion source; and the ion formed pass through accelerating silts and become an unresolved ion beam. In the magnetic a field the ion travel in deflected circular paths, where radii are functions of the ion mass : charge ratio. The resolved beam is focused on the detector and the discharge current is amplified and subsequently recorded.

During acceleration through a potential difference V a kinetic energy ev is given to the ion. All ions having same charge are accelerated to the same energy level. In a magnetic field with a direction of induction normal to the direction of motion, the ion beam will be deflected in a circular path.

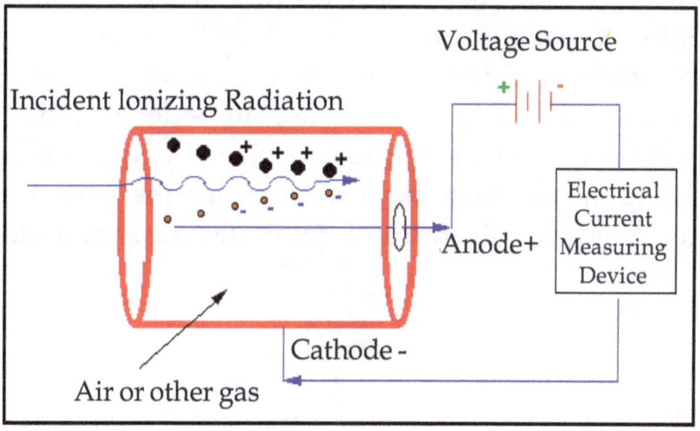

Fig. 8.16: Gas filled detector

The commonly used equation to describe the interaction is

$$r = \frac{144}{\beta}\frac{\sqrt{mV}}{\eta}$$

Where, r is the radius of deflection in centimeters, m is mass of an ion in daltons. V is the voltage applied to the electrode of an ion source. β is the magnetic induction in gasses, and η is the electric charge of the ion.

The Mass Spectrograph

The mass differences that arise in nuclear processes, which are typically of the order of several parts in 10,000, are not very large, but they can be detected by precision measuring equipment. We begin with a brief survey of the principles of operation of such an apparatus, the mass spectrograph.

In mass spectrograph an ion source produces positive ions of the species whose mass m is to be determined. The ions (each carrying a charge q) emerge from the source at low velocity and are accelerated towards the silt by the large accelerating voltage. The kinetic energy of the ions on reaching the silt is equal to the work done on then by the accelerating voltage.

$$\frac{1}{2}mV^2 = q^r$$

On passing through the silt, the ion enters a magnetic field depicted into the plant of the paper and continues in a circular path at constant speed, striking a photographic plate (or other suitable detector) and leaving a mark when the plate is developed. The radius of the path of ion in the magnetic field is given by equation, combining the two equation, we obtained a formula for the mass of the ions. Formula for ionic mass in form of quantities measured on the mass spectrograph.

$$r = \frac{m\gamma}{q^{\beta}}$$

β = magnetic field
q = ion charge
r = measured path radius
γ = accelerating voltage.

8.2.2.2 Non-pulse systems

Non-pulse systems can be divided into current producing system in which the output is proportional to the total ionization current produced within the detector and the photographic system.

I. Photographic system

When a piece of film is exposed to radiation, the emulsion reacts in the same way as when struck by visible light. The amount of radiation can be estimated from darkening of the film. Film can be considered a nuclear detecting device, it is used either in the form of a film badge to determine the accumulation of radiation or as a data-registering device a localizing detector as with auto radiogram. The components of any film or plate are

(1) The sensitive silver halide material.
(2) A dispersal medium for the sensitive material usually gelatin, and
(3) The emulsion base, which is cellulose acetate for most films and glass for slides.

A. Photographic Emulsion

A photographic plate seems as the detector in the accidental discovery of radio activities. The type of ionization process that results when photosensitive material is exposed to radiation blackens the emulsion when the material is developed. This method of detection is still in widespread use. The badges worn by workers in radiation areas are plates of photographic film which are periodically developed to measure the amount of the wearer(s) exposure to radiation during the period. Photographic plates are also used in various research projects to record the passage of radiation and ionization particles.

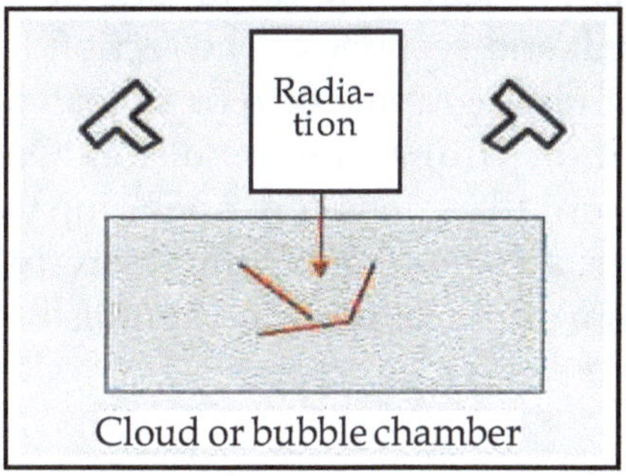

Fig. 8.17(a): Photographing the particle tracks

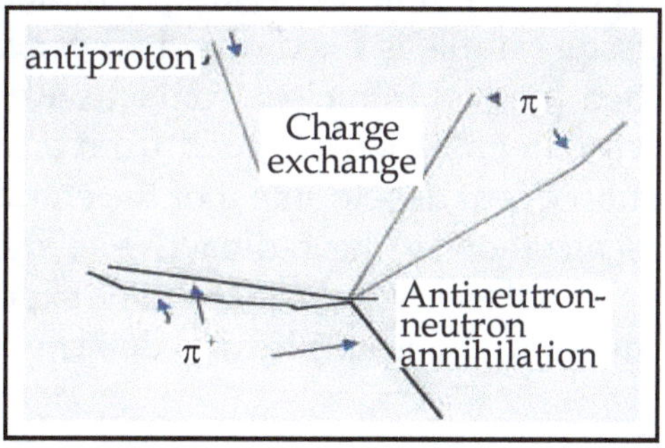

Fig. 8.17(b): A sketch of the tracks of change exchange and antineutron-proton annihilation

B. Film Badge

For personnel monitoring, the density of the blackening on the emulsion may be compared with that of a calibrated film.

II. Current producing system

Measurements are made with ionization chambers with an integrating circuit. The chambers themselves are made in a variety of forms and one used for measuring gaseous or solid samples. They are connected to either a vacuum tube or a dynamic capacitor electrometer circuit. The electrometer gives reading by either the rate of drift or the constant detection method. This technique is particularly useful in the measurement of C^{14} in small bone samples, where the tissue can be dissolved and counted in the form of CO_2 or a solid carbonate.

A. Autoradiography

The oldest method used for recording and crudely measuring radioactivity was that of fogging of photographic emulsions, used for this purpose as early as 1896 by Becquerel. This procedure has now been greatly reined to the extent that it may be used under certain conditions for quantitative measurements in addition to the usual qualitative demonstration of radioactivity by auto radiograms.

In its simplest terms, an auto radiogram is the image produced on a photographic emulsion by radioactivity emanating from the sample. Many variations in procedure and in materials are available, and the choice depends upon the conditions of the observations to be made and upon the kind of records it is desired to produce. In general, following administration of a radioactive element to an organism, a tissue sample is fixed, dried embedded, and sectioned. The section is then pressed into close contact with a photographic film or is coated with an emulsion stripped from a film or with a fluid emulsion. On subsequent development of the emulsion, all regions corresponding to localization of the radioactive element are darkened; prints may be enlarged to any desired size. The exposure time for the film or emulsion must usually is empirically determined.

B. Temporary contact auto radiography

With this method, an emulsion mounted on a film base is placed in firm contact with the specimen for a predetermined exposure time. After exposure, the specimen and film are separated and the film is processed to reveal the distribution of radioactivity in the specimen. Usually, X-ray film (non screen or dental X-ray film) is used. The temporary contact method is particularly useful for gross anatomic sections; histological sections mounted on glass sides, and paper and thin layer chromatograms. Compared to permanent contact techniques, temporary contact autoradiography furnishes poor resolution, because intimate contact is difficult to achieve between the emulsion and all portions of the specimen.

C. Permanent contact autoradiography

In improve resolution and reduce problem of alignment, histological sections containing radioactive material may be maintained permanently in contact with a photographic emulsion. Usually, nuclear emulsion such at those in Eastman Kodak's ITA or NTB series are used for permanent contact autoradiography. Compared to emulsion on X-ray film, nuclear emulsions have a greater concentration of silver bromide granules of small and very uniform size. Consequently, these emulsions provide high resolution and a reasonable sensitivity to β particles. After exposure, the emulsion is developed to precipitate metallic silver directly on the section, and the deposited silver is viewed simultaneously with the section by either bright field or phase-contrast microscopy. The distribution of radioactivity in the specimen may be estimated quantitatively by counting silver grains in various, regions of the specimen. A variety of techniques have been developed for permanent contact autoradiography.

Electron microscopic autoradiography

Relatively recently, techniques have been developed for electron microscopic autoradiography at very high levels of magnification. An ultra thin specimen embedded in plastic is coated with a very fine grain emulsion (grain diameters of 0.03-0.1 μ) by dipping a fine copper grid containing the specimen into liquid emulsion. Alternately, a metal 100 P may be dipped into liquid emulsion and the captured emulsion deposited over the grid containing the specimen. A resolution of 0.1-0.3 μ is obtainable by electron microscopic autoradiography, some times referred to "molecular autoradiography".

Auto radiographic Resolution and Artifact

The resolution furnished by an auto radiographic technique may be described as the minimum distance required between two point sources or radioactivity for the source to be resolved as two distinct images in the autoradiography. In general, the resolution is improved as the following factors are reduced:

(1) Emulsion thickness

(2) Grain size

(3) Specimen thickness

(4) Distance between specimen and emulsion and

(5) Range of the radiation.

The estimated mean ranges of β particles in a photographic emulsion are listed in Table-8.1 for a few nuclides of biologic interest. Tritiman furnishes auto radiographs of superior resolution; higher energy β particles penetrate the emulsion to greater depths and provide images which are less distinct and less useful for studies such as intra cellular localization. Similarly, X-ray and γ rays usually provide auto radiographic images of inferior quality, although soft x-rays such as those from ^{55}Fe can produce images of acceptable quality when used with special emulsions. Because higher specific activities can be obtained by labelling compounds with ^{3}H rather than ^{14}C, and because ^{3}H provides superior resolution, this nuclide is preferred for most auto radiographic studies.

Auto radiographic images are susceptible to a variety of artifacts related to inappropriate care of the specimen or emulsion. For example, the specimen must be stained and dehydrated carefully to prevent the radioactive material from leaching from the specimen or changing its distribution within the specimen. For specimens coated with an emulsion prior to staining, the staining process should not include materials which are absorbed by the emulsion for example; hematoxylin-cosin stain is absorbed strongly by many nuclear emulsions. Artifices in the auto radiographic image may be caused by conditions such as volatile agents, in the specimen, mechanical pressure on the emulsion, the presence of light, dust and debris during exposure and processing of the autoradiograph, and expansion or contraction of the specimen or emulsion during exposal and processing. The presence of some artifacts may be revealed by preparing duplicate "autoradiograph" for specimens which contain no radioactivity.

Tables 8.1. Estimated mean Range of β particles in a photographic emulsion

Nuclide	Average β energy Mev.	Range, μ
^{3}H	0.0055	1
^{14}C and ^{15}C	0.049	10
^{131}I	0.188	100
^{32}P	0.70	800

Difficulties in autoradiography

The four main differentials in autoradiography are (1) artifacts,(2) movements of the tracer after completion of the experiment, (3) leaching of the tracer, and (4) in fine micro autoradiography the presence of unit near ported tracer. These problems are usually overcome with careful work and attention to detail.

Artifacts, that are some sort of image or impression on the emulsion that is not due to radiation, may arise through excessive pressure or chemical reaction of the plant with the emulsion.

Artifacts due to pressure or plant secretions of a chemically active nature are easily checked by the simple means of putting a number of specimens untreated with radioisotope through the identical processes as the at routed material. The possibility of chemical action can also be prevented by placing a piece of very thin "Saran warap", "Mylar" melinex or similar plastic sheet between the specimen and the emulsion.

The problem of tracer movement during drying and exposure of the film is harder to solve. Most plants are prepared for autoradiography by drying in the oven or freeze drying, although sometimes fresh material is applied direct to the film and placed in a deep freezer for exposure. Before any form of drying, plants are usually mounted on herbarium drying paper or blotting paper using a few strips of "scotch" tape to hold them in an appropriate position. It is then easy to prevent gross movement of the radioisotopes between root and stem or between stem and leaves, simply by using a razor blade to cut through each petiole or leaf base, and through the base of the stem, thus isolating all the major parts and preventing movement from one to another during drying. Movement within broad dicotyledonous leaves, which can be significant, can not be controlled in this manner, and the only answer is freezing or freeze drying.

The leaching of tracer and the presence of unincorporated tracer are not normal problems with whole plants, but can be significant in the autoradiography of thin tissue sections for the study of transport of molecules within cells by the "pulse chase" method, and in the localization of metabolite synthesis.

Handing errors common to any careless photographic processing may also occurs scratching, finger prints, stray light, and non uniform development, displacement of the film during exposure and extraneous dust or other debris. An effect of histological strains on the emulsion can occur with micro autoradiography.

Neutron moisture gauges

The principle of the neutron moisture gauge was indicated by the early work of Chadwick (1932), who discovered neutrons through their interaction with materials rich in hydrogen. He noticed that when alpha particles collided with a piece of beryllium in the presence of paraffin, protons were detected. In the absence of paraffin, protons were not detected. From this and later experiments, it was found that neutrons are uncharged particles of mass slightly larger than that of a proton, and that they interact with soil and other materials primarily in two ways scattering and capture. Scattering occurs when a neutron collides with a nucleus of atom and transfers some or all of its kinetic energy to the nucleus. The scattering process continues until the kinetic energy of the neutron approaches that of the surrounding atoms. Capture (or absorption) occurs when a neutron enters the nucleus of the atom and ends its existence through a nuclear reaction. Depending upon the kinetic energy of the neutron as well as the kind of atom penetrated, an atom of a different element will be formed with the emission of a proton or an alpha particle, as well as gamma radiation. The utility of a neutron moisture gauge is based upon the physical laws governing neutron scattering and absorption.

Fig. 8.18: Neutron moisture gauge (Neutron probe)

Principle of method

Neutrons emitted from a sources used in neutron moisture gauges have energies changing from near zero to 10 MeV and are classified as fast neutrons. Neutron scattering reduces their energies to those in thermal equilibrium with the constituent atoms of the soil and such neutrons are classified as slow neutrons. Slow neutrons are further classified into thermal neutrons (energies up to about 0.5 eV) and epithermal neutrons (from 0.5 to about 2 eV).

When a source of fast neutrons is placed in a soil, the neutrons collide with atomic nuclei of the soil and are scattered in all directions. With each collision, a neutron not only losses some of its energy but changes its direction of movement. The average energy loss by a fast neutron is greater for collisions with atoms of low atomic weight than for collisions with heavier atoms. Hydrogen, a component of water, is the most effective neutron scatter in the soil. Hence, the greater the soil water content, the greater is the opportunity for fast neutrons to be reduced to slow neutrons in the vicinity of their source. Once neutrons have slowed to thermal energies, they diffuse through the soil very much like one gas diffuses except that their lifetime and spatial distribution are limited by capture (absorption). A source of fast neutrons placed in a moist soil immediately becomes surrounded by a cloud of thermal neutrons. The density of this cloud represents equilibrium between the rate of emission of fast neutrons, the rate of their scattering or thermalization and the rate of their capture. A slow neutron detector insensitive to fast neutrons placed in the vicinity of the fast neutron source allows a relationship to be established between the soil water content and the number of slow neutrons in the cloud.

❑❑❑

Chapter 9

Difficulties & Their Correction in Radio Tracer Assay

9.1 Potential Tracer Difficulties and their Corrections

In radioactive measurements, the observed counting rate is not equal to the true disintegrations rate of the sample since a number of factors like nature of detectors tend to increase it or, in a majority of cases, decrease the actual counts. Some of the important factors responsible for this are

1. Background,
2. Backscattering
3. Absorption
4. Self absorption
6. Geometry
7. Instrument efficiency

9.1.1 Background

The counts registered in the counter are not only due to the disintegrations of the sample but also due to those contributed by

1. Cosmic radiation
2. Natural radioactivity in the vicinity

3. Artificial radioactivity due to fallout or an X-ray Unit nearby
4. Thermal noise, etc.

These extra counts due to the above factors are called Background Counts and can be minimized by avoiding material hazards through the adoption of various clean methods, proper shielding techniques against fallouts and X-rays, anticoincidence circuitry, etc. in the counting systems. In spite of the above, background counts due to certain other causes are inevitable and hence necessary corrections have to be made for the same for obtaining the true counts.

Actual counts (C_a) = Observed counts (C_o) - Background counts (C_b). The net counts obtained after deducting the background is not also the actual count as it is modified by the amount of back scattering, adsorption, self absorption, geometry, coincidence loss and instrument efficiency.

9.1.2 Back scattering

Samples emit radiations in all directions, some of them towards the detector and some of them away from it. Part of the radiations emitted away from the detector are scattered back towards it and this has a tendency to increase the counting rate. This phenomenon is called Back Scattering.

The back scattering factor (f_b) is determined by first counting the sample on nylar film and later measuring the same with varying thickness of backing materials. It will be found that the back scattering count rate increase proportionately up to a certain thickness above which it remains constant. This value is called the Saturation Value and the minimum thickness of the backing material required to reach the saturation value is called the Saturation Back Scatter Thickness for the emitter. This back scatter thickness varies with the material. The ultimate back scatter thickness value (f_b) attained at saturation thickness is independent of β-energy within a range of 0.3 Mev and 2.3 Mev for a given backing material. The saturation value at the saturation back scattering thickness is directly related to the atomic number of the backing material; f_b values for lead and platinum are about 1.8; whereas for aluminum it is 1.30. In view of this, planchets of the same material and of the same saturation back scattering thickness should be used throughout the experiment to avoid variation in the values. Further, a planchet designed to give a saturation thickness for one radioisotope, say ^{14}C could be too thin for another isotope ^{32}P, and even a slight variation in thickness from planchet to planchet would produce a significant difference in counting rates.

In an addition to the back scattering, Side Scattering of radiations can occur to a certain extent and this can increase the observed counting rate noticeably if materials of high atomic numbers are used near the source of the detector. Hence sample holders, usually of leucite or other low atomic numbers, are used.

9.1.3 Absorption

The radiation emitted could be absorbed by the air, window of the tube or by the sample itself. The absorber window of the tube can be considerably reduced by the use of extra thin aluminized nylar window of about 0.15 mg/cm^2 thickness as at this thickness the absorption is negligible. Also, considerable gas flow is maintained through the counter to prevent any leak. Air absorption is very little for α-rays whereas (f_w) is determined by means of an air absorption curve prepared with various thicknesses of external aluminum absorbers. In the case of thick samples the radiations, particularly β-particles from the lower layers, are absorbed by the layers in the overlying sample material. As the thickness of the sample is increased, the radiations from the topmost layers alone are measured, thus reaching a saturation thickness. Further, if the specific activities of the samples are varied, it will be found that the observed activity of saturation thickness is directly proportional to the respective specific activities. When a constant total activity is maintained in a series of samples, each diluted with increasing amounts of unlabelled material.

The phenomenon of saturation thickness will not be attained as the specific activity decreases or as the total count of the sample increases.

9.1.4 Self absorption

Self absorption is of little importance with Y-emitters as they are transmitted through the thickest assay sample with practically no self absorption. Similar is the case with high energy β-emitters as the sample thickness is usually less than 15 mg/cm^2. But the self absorption is of importance in the case of low energy β-emitters such as ^3H, ^{35}S, ^{45}Ca. Alpha emitters, because of the typically short ranges, are always measured at the saturation thickness. Also, the liquid samples emitting β -particle are adjusted so that the saturation thickness is exceeded.

Correlations due to self absorption effect are carried out with the help of the self absorption graph drawn by using a series of samples of constant specific activity but varying thickness. The apparent specific activity is plotted against sample thickness in mg/cm^2 (or weight of the sample if a standard planchet is used). It is then extrapolated to zero thickness and the specific activity at the zero point is the true specific activity of the sample (using the semi long paper). A second curve is then prepared of the percentage specific activity (apparent specific activity/true specific activity), that is, percentage transmission Vs sample thickness, taking the true specific activity as 100 per cent transmission. The correction factor for any sample thickness is read from the curve. The apparent counting rate divided by the percentage transmission corresponding to the sample thickness in question will then give the counting rates corrected to zero thickness or 100 per cent transmission. However, the curve has to be prepared for each instrument and for each set of assembly.

9.1.5 Geometry

The other factor which contributes to variation in counting is the mission of radiation in all directions due to the placement of the detector with respect to the sample. Of the total volume of radiations, only a fraction reaches the sensitive volume of the detector. As the sample is spread over a considerable area, it becomes an extended source and it is difficult to calculate the solid angle for the extended source. So the size of the source should be as small as possible and, for comparative measurements, counting has to be done in samples of uniform size and distribution: and also, it should be placed at the same distance from the detector. For a point source placed at the centre of a planchet, the solid angle will be equal to

$$2n = \left[1 - \frac{d}{d_1^2 - r^2}\right]$$

where d_1 is the distance of the source to the detector and r is the radius of the detector.

9.1.6 Instruments efficiency

Another important factor which modifies the total counting is the efficiency of the instrument which varies from one instrument to another and also with changes in temperature, pressure, voltage, line ageing of the tube, etc. Variations due to the instrument could be controlled by using the same instrument throughout but variations due to the other factors will be extremely difficult to control and determine. Hence prepared standards of radioactive substances of long half lives are used for normalizing the activities every day.

9.1.7 Coincidence loss

Another factor which contributes to variation in counting is the capacity of the instrument to resolve two incident radiations. A radiation detector requires a finite time to clear each pulse created by each radiation. If two or more radiations occur at the same time the detector will be able to account for only one and thus there can be Coincidence Loss. This loss could be due to amplifier circuitry but this is negligible as the resolving time has been reduced to 5 μ sec. Another source is the register used for recording counts and even there the loss could be reduced by the use of registers with high scaling factor having a resolving time of 1 μ sec.

The most important source of the loss is in the G.M. counters which have a resolving time of 100 to 300 μ sec. The coincidence loss is not normally high in small number of counts per second but this becomes important when the counts are more than 2000 per second. In such cases, one of the methods to bring down the losses is to dilute the sample and reduce the counts per unit time.

As a result of coincidence loss, the observed count is always less than the true count. If the resolving time is around 200 μ sec, an approximate correction of 0.5 per cent per 1000 cpm can be added to the observed count to give the true count. For accurate work, if the resolving power is known, the correction can be calculated using the formula $N_1/N_0 = 1(1\text{-}N_0T)$ where N_1 = true count, N_0 = observed count and T = resolving time. If the resolving time is not given, the same can be calculated using paired source methods wherein the counts from two sources are measured individually and also together. Normally, combined counts should be equal to the sum of the individual counts but due to the dead time or coincidence loss, it will not be so. From this the dead time or resolving power, (T) can be calculated using the formula $T = \frac{n_1 + n_2 - n_{1,2}}{2n_{1,2}\, n_2}$ and from this the correction due to the coincidence loss can be calculated.

Hence in making comparative measurements of β-activity with G.M. countries, corrections for all the factors may be supplied wherever necessary and true counts obtained. But in the case of samples emitting Y-rays, all factors in the equation may be constant while in liquid scintillation, the value of the factors may be standardized for comparative purpose or ignored.

In spite of making corrections for the above processes if repeated determination of activity of the same radio nuclide is made for the same duration of time, the results will be varying over a wide range of values due to radioactive decay being a random process. If the data are plotted in a graph, it will be in the form of normal distribution, and for most radioactive measurements, it will be a close approximation of the poison distribution curve. The mean value is the best estimate of the true value and the standard deviation gives a measure of the deviation ($\sigma = \sqrt{n}$) *where* n= total counts. In tracer work, if it is not possible to make larger number of measurements, usually only one or two measurements per sample are made. Because of the nature of poison distribution, the standard deviation of a single determination can be taken as $\sqrt{n}\,(\sqrt{number\ of\ counts})$. The larger the number of the number of counts collected, the smaller is the relative standard deviation. Radio assay is usually made to an accuracy equivalent, to 1 per cent standard deviation.

The standard deviation calculated is the combination of errors from sample and background counts. When the ratio of gross sample counting rate to background rate is high (over 10 times), the σ of the net sample radioactivity can be taken as the square root of the gross sample count. As it becomes smaller, the contributions of the background error to the error of the gross sample counts become significant.

When the sample is of low activity, for example, if the count rate of the sample is four times that of the background only, it is necessary to count the sample and background over a definite duration of time to that the relative standard deviation of each of the two operations is of a complementary magnitude. To this end the ratio, of background counting time t_b to that of the sample, is related to the ratio of the counting rates of the background n_b and the sample, n, that is, $t_s/t_b = \sqrt{n/n_b}$

Obviously, when the ratio n_s/n_b approaches unity, in order to realize a respective relative of the net counting rate one must count the sample as well as the background, over long period of time. To avoid lengthy counting operations it is essential that the counting rate of the final sample is considerably higher than that of the background (say 10 times). Also, it is essential to avoid contamination of the laboratory equipment, which raises the background rate and hence lengthens the required counting time.

9.2 Basic Difficulties in Tracer Methodology

9.2.1 Radio colloids

Some of the factors tending to promote colloid formation are as follows:

(a) Use of solvents in which the tracer tends to hydrolyze or form an insoluble compound,

(b) Presence of foreign particles in the solution,

(c) Presence of certain electrolytes and

(d) Increased age of solution

9.2.2 Parent –Daughter Relationships

(a) *Shorter-lived parent:* If the parent is shorter-lived than the daughter, then no equilibrium is reached between the parent and daughter. Starting with the parent activity initially free of the daughter, as the parent decays, the amount of daughter will increase and eventually decay with the half-life of the daughter.

(b) *Longer-lived parent:* If the parent is longer-lived than the daughter, it becomes possible to use the measured amount of daughter present in a sample as an indication of the amount of parent present. The important point is that, starting with the parent activity, equilibrium will be reached after a time equal to about 6 to 10 daughter half lives. At equilibrium the daughter will be decaying at the same rate of which it is formed, and therefore the amount present will be proportional to the amount of parent present. Also the apparent half-life of the daughter will be the same as that of the parent. When the parent is being constantly replenished, as in pile bombardment, or is very much longer-lived than the daughter, the term secular equilibrium is employed. In this case the parent does not decay measurably during many daughter half-lives. When the parent is relatively little-longer-lived than the daughter, the term, transient equilibrium is used. Clearly, secular equilibrium is the limiting case of transient equilibrium, and whether or not the equations for secular equilibrium can be applied depends upon the accuracy desired and the particular experiment.

In many cases the radiation characteristics of the short-lived daughter make radio assay of the daughter much easier than measurement of the parent. Also, when chemical separations of the daughter are convenient, it may lead to analytical methods that have the following advantages:

(1) The radioactivity present in the chemical separated fraction represents only that from the parent activity and therefore, the results are unaffected by any radio-contaminants in the original sample which are not carried through with the daughter fraction,

(2) The separated daughter activity is usually present in a smaller amount of extraneous mass and can therefore be measured with less self-absorption losses than when it is measured in the original sample,

(3) The radiation characteristics of the daughter serve to identify the parent,

(4) The original sample can be analyzed many times by allowing time for equilibrium to be reached between successive chemical separations of the daughter. The chemical separation and radio assay of Y^{90} from the samples containing Sr^{90} - Y^{90} is a good illustration of the advantages of this procedure. This method also allows differentiation between Sr^{90} and Sr^{89}, the latter having no radioactive yttrium daughter.

In a few cases a process known as isomeric transition (abbreviated IT) occurs in which a radioactive daughter is isotopic with the parent.

9.3 Summary and Suggested Working Rules

General requirements for radiation control may be stated as follows:

(a) Knowledge of the amount and nature of the radioactivity received,

(b) Under-standing of the chemical and physical behavior of the radioactive material in the experiment to be performed,

(c) Repeated performance of "dummy runs" to eliminate unforeseen difficulties

(d) Forethought in planning of operations

(e) Proper use of suitable instruments; and

(f) Common sense, vigilance, good housekeeping, and good work habits.

9.4 Radioactive Contamination of Soil and Plants

During the past fifteen years new components, the radioactive isotopes artificially synthesized by man, have appeared in the biosphere of our plant. Thus, new radioactive emitters were added to the natural radioactive elements scattered in nature. This contamination of biosphere is associated with the rapid growth of atomic industry and the increased use of nuclear energy.

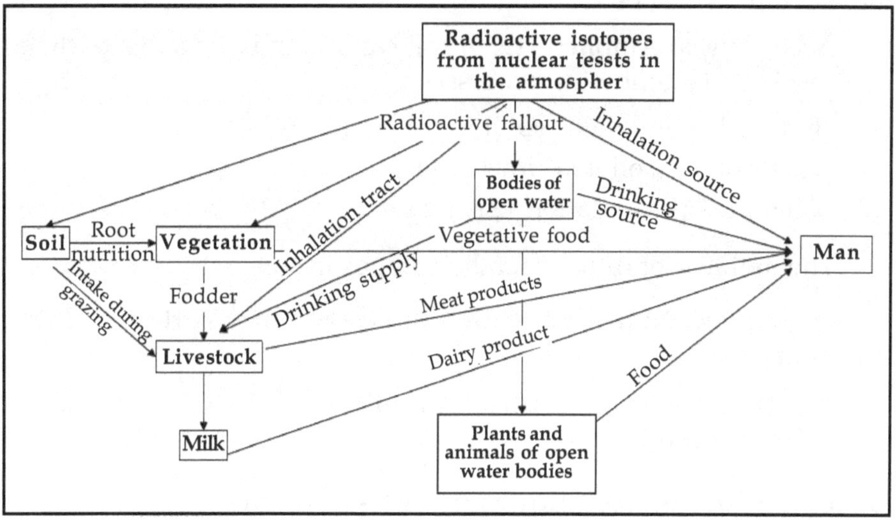

Fig. 9.1. Migration digaram of radioactive fission products from the external medium into man.

The problem of the radioactive contamination of biosphere has acquired special urgency since the detonation of atomic and thermonuclear weapons have caused the global contamination of our plant with radioactive substances and world wide intensification of the Y and β radiation background. Radioactive fission fragments of u^{235} and Pu^{239}, the main nuclear fuels, are released into the atmospheric and slowly settle on the earth's surface together with atmospheric precipitations. Radioactive fission products of heavy nucleus which are formed in nuclear explosions and which settle on the ground together with rainfall, have acquired a special term radioactive fallow.

The problem of the radioactive contamination of the biosphere has attracted the attention of a wide circle of specialists: physicians, biophysicists, pedologists, and geophysicists, since it become evident that ionizing radiations, whose source is the radioactive isotopes scattered in nature, have a possible dangerous action with grave consequences on living organism. An extensive study was begun on the migration of radioisotopes in the biosphere along the biological food chain, soil-plant-animal-man.

It has been recognized as an obvious fact that the principal danger produced by nuclear explosions is associated with the ingestion by man of farm products contaminated by radioactive substances. The use of food of plants thus contaminated is the most direct was that these products can be entering man.

According to Amphett (1955) soil can retain radioactive particles as a result of the following processes:

1. Ion exchange (water soluble fraction Sr^{90}-Cs^{137})
2. Adsorption (colloidal fractions)
3. Chemical precipitation (Zr^{95})

The mobility of different radioisotopes in soil :

For sorption: Nutherium >Strontium> Cerium > Yttrtum > Cobalt > Cesium,

For desorption: Strontium>> nutherium> Cerium > Cobalt>> Cesium > Yttrium.

Decontamination techniques in soil and plants

1. *Plowing factor:*
   e.g. Sr^{90} distribution in soil.

Depth (cm)	Untouched (%)	Plowing
0-5	79	31-57
5-10	13	29-42
15-30	08	14-27.

2. Imbedding the fission products below rhizosphere.
3. Application of lime and compost.
4. Potassium fertilization decreases Sr^{90} and G^{137} uptake.
5. Sr^{90}/Ca ratio in plants may be reduced by soil application C_a-carriers.
6. One may expect a decrease in Sr^{90} uptake by adding stable Sr-compound to soil, radio strontium will be diluted with stable isotopes.

❑❑❑

Chapter 10

Application of Radio Isotopes in Different Fields

Radionuclides are used in two major ways: for their chemical properties and as sources of radiation.

Radionuclides of familiar elements such as carbon can serve as tracers because they are chemically very similar to the non-radioactive nuclides, so most chemical, biological, and ecological processes treat them in a near identical way. One can then examine the result with a radiation detector, such as a Geiger counter, to determine where the provided atoms ended up. For example, one might culture plants in an environment in which the carbon dioxide contained radioactive carbon; then the parts of the plant that had laid down atmospheric carbon would be radioactive.

In *medicine*, radioisotopes are used for diagnosis, treatment, and research. Radioactive chemical tracers emitting gamma rays that can provide diagnostic information about a person's internal anatomy and the functioning of specific organs. This is used in some forms of tomography single emission computed tomography and positron emission tomography scanning. Radioisotopes are also a promising method of treatment in hemopoietic forms of tumors, while the success for treatment of solid tumors has been limited so far. More powerful gamma sources are used to *sterilize* syringes and other medical equipment. About one in two people in Western countries are likely to experience the benefits of *nuclear medicine* in their lifetime.

In *biochemistry* and *genetics*, radionuclides are used to label molecules and allow tracing chemical and physiological processes occurring in living organisms, such as *DNA replication* or *amino acid* transport.

In *food preservation*, radiation is used to stop the sprouting of root crops after harvesting, to kill parasites and pests, and to control the ripening of stored fruit and vegetables.

In *agriculture* and *animal husbandry*, radionuclides also play an important role. They are used to produce high intake of crops, disease and weather resistant varieties of crops, to study how fertilizers and insecticides work, and to improve the production and health of domestic animals.

Industrially, and in *mining*, radionuclides are used to examine welds, to detect leaks, to study the rate of wear of metals, and for on-stream analysis of a wide range of minerals and fuels.

Most household *smoke detectors* contain the radionuclide *americium* formed in *nuclear reactors*, saving many lives.

Environmentally, radionuclides are used to trace and analyze pollutants, to study the movement of surface water, and to measure water runoffs from rain and snow, as well as the flow rates of streams and rivers.

Natural radionuclides can be used in *archaeology* and in *paleontology* to measure ages. When radioactive carbon, for example, is in the atmosphere, it rapidly becomes separated from its decay products. Once it is bound up in a solid, such as wood or paper, its decay products must remain in place. So by measuring how much of these decay products has accumulated, one can estimate the time when the carbon was captured into solid form.

10.1 Radioisotopes in Agriculture

Less than 1/3rd of our planet is occupied by land but about 98 % of the food produced in this planet comes from this land. Land is the most important natural resource of this planet. Proper management of land is the almost necessary for the survival of the humanity. The decline of soil fertility had been the cause of fall of great civilization of the world in the past and it has become again the biggest threat to our environment. Our greed, growth and developmental activities are mainly responsible for the overall environmental degradation.

The US Academy of Science and the United Kingdom's Royal Society recently warned that "the future of our planet is in the balance. Sustainable development can be achieved, but only if irreversible degradation of the environment can be halted in time". Sustainable development in agriculture and industry is achievable only when the approach is ecologically, socially and economically viable. Nuclear and related isotope and radiation techniques are an important part of such integrated approach in which nuclear techniques compliment conventional or classical techniques in research and they provide data which other methods cannot generate.

The new challenges in Indian agriculture are to sustain the benefits of green revolution in the irrigated belt of the country and to user the second green revolution in the dry land agriculture. The progress in science and technology with special reference to nuclear technology over the last few decades made significant contribution in the growth and development of the mankind around the world. The usefulness and application of nuclear and related techniques in facing the new challenges, has been convincingly demonstrated in developing and developed countries. The effectiveness and utility of nuclear allied techniques in food security, sustainable agriculture, resources management, climate change and protection of environment has already been shown by our scientists in India.

To face the new challenges in Indian agriculture, radiation and isotope techniques are going to play very important role in the development of new agrotechnology which are otherwise not possible by conventional techniques.

The significant advances made in India or elsewhere using nuclear and isotope techniques to improve the quantity and quality of agricultural produce, reduce the post harvest losses, improve soil fertility, fertilizer and water use efficiency, increase crop production, control of pests and diseases and to protect the environment. Spectacular progress has been made in genetic improvement of crops through combined use of induced mutation and various other techniques.

10.1.1 Soil Fertility and Productivity

Indian scientists have done remarkable work in this field as India was the pioneer in the use of isotopes and radiation in agriculture at field level. The use of ^{32}P and ^{15}N helped the scientists to evaluate the performance of various phosphatic and nitrogenous fertilizers in different soils under different management conditions. The use of stable and radioisotope fertilizer management enabled to determine the correct dose, time and methods of application of fertilizers to the crops. It was possible to prepare the balance sheet of fertilizer nutrient in a particular cropping sequence. Including placement, time of application and fertilizer source have been studied e.g. for rice and maize and many other crops such factors have a direct connection with N transformation and possible N losses and in this area detailed studies have been connected with:

- Soil conditions and placement which encourage direct volatilization of NH_3-Nitrogen.
- NO_3-Nitrogen losses by leaching, de-nitrification and other chemical or biological processes.
- The relative amounts of soil and fertilizer nitrogen in the pool of available nitrogen.
- Symbiotic nitrogen fixation by legumes in the field.
- Fixation of NH_4-nitrogen on clays.
- The degree of biological activity and immobilization.
- Transformation of N.
- The available nitrogen status of soil ('n' value)
- Recovery of fertilizer-N.
- Balance sheet and residual effects

Efficiency of N-sources studies involving very small amounts of N and short uptake time (5-120 min) have been carried out comparing the uptake of NO_3 and NH_4 by excised roots and by seedling we may except that the development of the emission spectrometer technique will encourage many biochemical and physiological studies of N-uptake, utilization, turnover and protein formation in plants.

From the balance sheet it was observed that N use efficiency has been found to be hardly 30-40 per cent for paddy crops under average farming conditions in India. Inefficient N utilization is largely due to N loss through ammonia volatilization and denitrification through various routes (Sinha, 2001). In case of wheat the fertilizer N utilization can go up to 60 per cent and out of the remaining 40 per cent hardly 2 to 4 per cent remains available to the next crop. So the residual effect of fertilizer N was found to be negligible .That is why it is essential to apply fertilizer N to every crop in order to harvest good yield (Deb, 1997).^{15}N balanced study revealed that application of prilled urea and FYM resulted in the higher retention of fertilizer N in soil., lower volatilization loss of ammonia from fertilizer source and lower recovery of fertilizer N by rice as compared to prilled urea alone (Jena et al., 2003). Jagadeeswaran et al., (2004) revealed that N derived from ^{15}N urea (Ndff) increased with number of split application of N up to four in turmeric rhizomes, both at 180 days (19.46 per cent) as well as at harvest stage (30.76 per cent). Increasing the splits beyond this declined Ndff.

The quantitative measurement of gaseous loss of N in field experiments under various management conditions reveal that fertilizer materials must be enriched with very high atom per cent excess (80 to 90 %) in order to get very reliable results. But such fertilizer materials are very expensive and every country cannot generate such data. Through Coordinated Research Programme of International Atomic Energy Agency, it is however, possible to generate such data in selected agro-ecological regions for certain cropping system. Through isotopic studies it is possible to show that nitrate fertilizers are more vulnerable to leaching losses under submerged conditions than urea and ammonium sulphate. Further, in order to reduce that leaching loss of N, split application of N was found to increase high fertilizer use efficiency and less leaching loss of fertilizer N. It has now become a recommended practice in wheat to apply one third N at basal and two-third after crown root initiation stage (about 20 days after germination) which is slightly superior to conventional practice of application 50% N at basal and 50 % N at crown root initiation stage. In rice crop 3 to 4 split application of N at different physiological stages of crop growth have been found to give the highest economic return from the recommended dose of N. Jha *et al.,* (2003) showed that only 40- 71 per cent nitrogen was derived from the fertilizer source. The highest per cent of nitrogen derived from fertilizer was observed in case of three split application of urea alone applied to wheat crop@ 120 kg N/ha (30+45+45 kg) before irrigation. Total nitrogen uptake varied from 40.3 to 76.3 kg/ha with dose of N varying from 90-120 kg urea- amended with DCU.

Legumes are known to fix atmospheric N but the amount of N fixed by different species could not be properly quantified. With the help of ^{15}N labelling techniques it has now become possible to quantify the amount of N fixed by different leguminous crops as well as by leguminous trees. Apart from legumes it has been made possible by ^{15}N isotope techniques to determine the extent of N fixed by certain non-leguminous plants such as blue-green algae and their associations particularly Azolla in lowland rice fields. Although Azolla has been used for centuries as a green manure for rice in China and Vietnam, but the reasons for increasing soil fertility through Azolla was not known till recently. With extensive studies using ^{15}N labelling techniques it was possible to demonstrate that 80-90% of the nitrogen in Azolla is derived from atmospheric N fixation. Consequently, Azolla did never compete with rice for native soil nitrogen. When Azolla is incorporated into the soil, it becomes an easily available N source like urea for rice. Further, residual effect of Azolla is more than that of urea to the next crop of wheat and wheat yield was found to increase remarkably in Azolla incorporated soils. Azolla has now being recognized as a very important biofertilizer for rice crop after ^{15}N studies showed its efficiency in fixing atmospheric nitrogen. The result from laboratory and green house study using radio tracer ^{86}Rb by Sud (2005) indicated that use of wheat or paddy crop residues were at par with FYM in terms of increased in yield, nutrient uptake and their recoveries by potato as well their residual effect on the succeeding garlic crop.

Retheesh *et al.*, (1997) use ^{14}C labeled urea to quantify N absorbed in the form of molecular urea. ^{15}N labeled and non-labeled materials were used to quantify N absorbed in the form of nitrate or ammonia. They observed that rice absorbed more nitrogen when supply with a combination of ammonium and nitrate in equal proportion. It also absorbed molecular urea in presence of other readily available forms of nitrogen which amounted to 12-18 percent of total N uptake. Uptake of urea reduced when supplied with ammonia and/ or nitrate.

Radioisotope studies with ^{14}C indicated that 4.07 per cent of the CO_2 fixed by loranthus leaves was translocated to the nearest coca leaf by the second day. This indicated that there is movement of prepared nutrients from the parasite to the host. The study also revealed that systemic herbicides can not be safely recommended for the control of loranthus (Girija *et al.*, 2006).

As indicated earlier, radioisotopes have been used widely in many fields including fundamental studies in soils. They have been used to estimate the contribution of nutrients from soil as well as fertilizer sources, phosphate fixation capacities, potassium release studies,

studies on soil organic matter, etc. To study the contribution of nutrients from soil and fertilizer sources, various methods have been proposed to estimate each one of the fractions. The available phosphorus has been calculated from the A, E and L values obtained using ^{32}P. The concept of E value was developed by Paneth and Vorwerk (1922) with thorium (an isotope of lead) to determine the amount of isotopically exchangeable lead in an insoluble lead salt, thus permitting the calculation of the specific surface of the salt. In this concept it is assumed that an exchange of ions results in the addition of an isotope of the same ion E* to the ions E that are in constant exchange with chemically identical ions in water, according to the following equation :

$$E\ (\text{surface}) + E^*\ (\text{solution}) \rightarrow E^*\ (\text{surface}) + E\ (\text{solution})$$

$$\frac{E^*surface}{E^*solution} = \frac{E\,surface}{E\,solution}$$

Based on this, Spinks and Barber (1946) pointed out that it is possible to measure the P taken up both from fertilizer and soil and also the utilization of the fertilizer P, something which could not possibly be done by ordinary chemical means. In 1948, McAuliffe *et al.* suggested that this principle could be applied to a soil system to measure the surface phosphate ^{31}P which they claimed to correspond to the level of phosphate fertility of the soil. They added a small volume of water containing a tracer of ^{32}P to avoid any substantial net movement of P either to or from the soil phase during the equilibrations.

$$^{32}P\ (\text{solution}) + {}^{31}P\ (\text{solid}) \rightarrow {}^{32}P\ (\text{solid}) + {}^{31}P\ (\text{solution})$$

$$^{31}P\,(\text{solid}) = \frac{^{32}P(\text{solid})}{^{31}P\,(\text{solution})} \times {}^{31}P\ (\text{solution})$$

In the above isotopic exchange of phosphates in soil, perfect equilibrium between added ^{32}P and the surface phosphate was not readily established within any reasonable period or equilibration probably due to the gradual diffusion of PO_4^- ions into the solid inorganic phosphate and/or due to the exchange involving PO_4^- ion other than true surface phosphates. The time of equilibration required was therefore selected in such a way that the contribution of slow exchange processes was negligible while the rapid exchange was completed within the period. Increasing soil solution ratio decreased the isotopically exchangeable phosphate but this effect got diminished at large equilibration periods. The influence of electrolytes, anions and temperature on the E values measured was found to vary with the properties of soils.

Later Friend and Dean (1952) introduced the concept that a plant having two sources of a nutrient in a soil will absorb from each in proportion to the respective quantities available. The quantity of nutrient drawn from the fertilizer source was measured by the use of isotopes using the formula

$$\text{A value} = \frac{B(1-Y)}{Y}$$

where A = Amount of nutrient available in the soil

B = Amount of nutrient applied through the fertilizer

Y = Proportion of the nutrient in the plant derived from the fertilizer source

The A values never warrant the attainment of an isotopic equilibrium between the added nutrient and the nutrient present in the soil, and is always measured in relation to standard fertilizer used.

Larseen (1952) was the first to suggest that when labeled phosphate fertilizer was added to a soil at different rates and plants were grown, specific activity analysis of plant material would give a constant value if the isotopic dilution of the added chemical was assumed to take place in the soil system. He assumed that the reason for the constant value obtained was that equilibrium has been reached between the added phosphate and the exchangeable phosphate in the soil. The isotopically exchangeable P of the soil was calculated using the formula

$$L = \left[\frac{C_0}{C_1} = 1\right] X$$

where C_0 and C_1 were the specific activities of the applied P and plant P respectively, X = the amount of phosphate added, and L = the quantity of soil P which had taken part in the isotopic dilution of the applied P. Larsen reported that L values were independent of the amount of P applied and also became independent of time after a particular period of plant growth, suggesting that the isotopic equilibrium was obtained. Semb *et al.* (1965) observed that there was a very close relationship between L value and P extracted by Olsen and Bray No. 1 extractants in samples having pH 6.3.

Russell *et al.* (1957) stated the E and L values were similar and could be regarded as alternate measures. The choice of these methods was dependent mainly of the convenience and economy of labour. Based on these E values were preferred. Fried (1964) considered L values as an E values in which the plant was being used as a means of sampling the soil solution upon equilibration of added P with soil surface P. He further suggested than in L value, the conditions of measurements were exactly similar to those of the plant grown in the soil. In the determination of soil P by isotopic dilution method, Larsen and Sutton (1963) started that ideally no P should be removed so that the equilibrium between labile and non-labile P is not disturbed. This condition could be obtained in E value determination but in L value measurements which involve growing of plants, removal of P was inevitable.

Fried (1964) and Larsen (1967) made clear that A value had no relationship to E or L values even though an identical formula was used in both A and L values. In the A value, the attainment of an isotopic equilibrium between the added nutrient and soil nutrient was not required as in the case of L value. Larsen and Gunary (1964) noticed that seed P gave an increased L value. To correct this, they estimated the amount of seed phosphate which had entered the harvested portion of the crop. The true specific activity of the phosphate taken up by the plant was arrived at by using.

$$S_P = \frac{S_S}{1-F}$$

where S_s = specific activity of phosphate in the plant material, F = fraction, S_P = corrected specific activity. The corrected specific activity of the plant P was taken in the L value equation. Thakur *et al.,* (2005) reported that 'A' value of phosphorus in soil increased at all the three growth stages of rice with successive levels of phosphorus application. Change in 'A' value become marked when P was added in combination FYM and PSB at every growth stages in various P status soil. It may described to the fact that low P gradient soil was more responsive to added P as compared medium and high P status soils.

Similarly, radioisotopes have also been used in a number of studies on phosphate fixation. In these studies, either phosphatic compound labeled with ^{32}P, or carrier free ^{32}P have been used.

Like nitrogenous fertilizers, the relative efficiency of phosphatic fertilizer was evaluated using ^{32}p in different soils. Results of a large number of experiments in cereals proved the superiority of DAP over superphosphate at the same level of N application to the crop. This resulted finally in the establishment of a large number of DAP factories in India and superphosphate factories were virtually closed. On an average, the amount of P derived from low, medium and high P fertility soils towards P absorption by wheat shoots was 43.0, 37.6 and 32.6 percent, respectively. Amongst P fertilizer sources, the relative effectiveness was follows: monocalcium orthophosphate(MCA) > diammoinum hydrogen orthophosphate(DAP)> single super phosphate(SSP)> dicalcium orthophosphate (DCP). On an average, DCP, DAP and SSP were 76, 93 and 85% effective as compared to MCP (Sharma *et al.,* 2004). They also reported that, there was no increasing capacity factor(b) values with increasing P availability of soils irrespective of level of applied P. Capacity factor of high P soil was 2.4 times higher than that of low P soils. Application of 40 mg/kg phosphatic fertilizers resulted in decreased in "b" value in all the soils. Magnitude of reduction in 'b' value was highest with MCP were as minimum decreased was observed with application of DCP.

It was also demonstrated using ^{32}p isotopic technique that in acid soils native rock phosphates of low p content can be used as phosphatic fertilizer without sacrificing any yield. To save sulphuric acid consumption, nitro phosphates of varying grades were manufactured in this country and tested their agronomic efficiency under field conditions. Results of ^{32}P labelled nitrophosphate studies strongly indicated that nitrophosphate must contain at least 50% of its total P_2O_5 content in water soluble form. The isotope ^{32}P being easily available and not expensive, it has been used extensively by Indian scientists to generate location specific research information with respect to P fertilizer use efficiency in relation to N , Zn, S and other fertilizer nutrients in different crops. Crop varieties were also tested for their P fertilizer use efficiency in relation to their root distribution pattern as determined by radioisotope technique. Our scientists have used ^{32}P placement technique for evaluation of root distribution pattern of fruit crops for increasing fertilizer use efficiency of horticultural crops especially in high density orchards.

Khatik *et al.,* (2004) found that maximum amount of applied P was transformed to Ca-P and minimum into salloid-P in all the plant growth stages of paddy. Different fractions of transformed P in the soil followed the order Ca-P>Fe-P>Al-P>Oculuded-P>salloid-P.

In mid sixties with the introduction of high yielding varieties of wheat and rice, inorganic fertilizer consumption increased tremendously in irrigated regions, increased biomass production by high analysis pure inorganic fertilizers depleted native zinc from soil at a very alarming rate. Consequently, widespread zinc deficiency appeared in wheat and rice crops all over the country. The response trial conducted by several researcher under pot culture and field condition using ^{65}Zn labeled zinc sulphate revealed not more than 2 per cent Zn is utilized by cereals crops (Deb,1997). Here again, through a CRP Project of IAEA, relative efficiency of different zinc fertilizers, their dose, time and method of application was studied in India in rice-rice and rice- wheat cropping sequence using ^{65}Zn labelled fertilizer under field conditions. This research helped the scientists to recommend application of 5 kg Zn/ha through zinc sulphate at the time of sowing to overcome the zinc deficiency in soil. The interaction of zinc with iron, manganese, nitrogen, phosphate and sulphate in their availability to plants and subsequent impact on crop yield was studied using radioisotopes of the respective nutrient elements.

Recently, scientists in India are experiencing widespread sulphur deficiency problems in different crops in 120 districts of the country.

Such problem appeared because of continuous use of sulphur free fertilizers in the last three decades. Using ^{32}S radioisotope, scientists have evaluated different sources of sulphur in this country. Results indicated that the efficiency of S depends on the levels of N and P fertilizers applied to the crops. The response of S fertilizers to crops was found maximum when superphosphate, ammonium sulphate or calcium sulphate was used as source of S. In oilseed crops, isotope aided studies revealed that superphosphate and gypsum are better than ammonium sulphate in increasing seed yield and oil content. The response to applied S was found maximum at the highest levels of recommended dose of N and P fertilizers to different crops. Soluble sulphate fertilizers when applied at the rate of 20 to 30 kg S/ha were give the maximum yield in cereals, oilseeds and pulses. Beyond this rate, the S use efficiency decreases drastically almost in all crops with little increase or decrease in grain/pod yield.

Apart from these, lot of other studies have been done by the scientists of India using appropriate stable and radioisotopes to rationalize the fertilizer schedule in different cropping systems.

Radioisotopes helped in the development of mechanistic models for predicting fertilizer uptake by plants under a given management condition based on soil and plant parameters. In such models the rate at which a nutrient moves to the plant root by concentration gradient (the diffusion coefficient) is the most important soil parameter which is required to be determined in the soils where plants are to be grown. This parameter is most conveniently determined by using radioisotopes of the nutrient to be tried. Our scientists have developed models for predicting the fertilizer uptake by plants under different soil conditions by using ^{32}P, ^{65}Zn, and ^{59}Fe. The results are very encouraging and our scientists are trying to improve their predictions by including all relevant parameters in the model. Since diffusion coefficient of an interactive nutrient expresses the integrated effect of prevailing soil conditions on the rate of movement of a nutrient to plant root, it very often indicates the bioavailability status of the nutrient under that particular soil condition. Lot of research information has been generated by our scientists in this regard which are no or less informative than the data generated in UK, USA, Australia and some other western countries in this regard. In all these studies, radioisotopes played the very important role.

The clay mineralogy of soil clays and fertilizer reaction products have been studied extensively using X-ray diffraction technique. A soil clay mineral map of India has been prepared indicating the geographical distribution of dominant clay minerals in soils of different agro climatic regions.

Table 10.1. Natural abundance of isotope which are used in agriculture

Element	Radioisotopes	Halflife	Stable isotopes	% Natural abundance
Hydrogen	^{3}H	12.26 year	^{2}H	0.02
Boron	^{12}B	0.02 sec	^{10}B	19.78
Nitrogen	^{13}N	9.96min	^{15}N	0.37
Carbon	^{14}C	5730 year	^{13}C	1.11
Oxygen	^{15}O	124 sec	^{18}O	0.20
Magnesium	^{27}Mg	9.5 min	^{26}Mg	11.17
Phosphorus	^{32}P	14.3 days	^{31}P	100.00
Sulphur	^{35}S	86.7 days	^{33}S	0.76
Chlorine	^{36}Cl	3 X 105 years	^{37}Cl	24.47
Potassium	^{42}K	12.4 hr	^{41}K	6.88
Calcium	^{45}Ca	163 days	^{42}Ca	0.64
Manganese	^{54}Mn	312 days	^{55}Mn	100.00
Iron	^{55}Fe ^{59}Fe	2.4 years 45 days	^{58}Fe	0.33
Copper	^{64}Cu	12.9 hr	^{65}Cu	30.91
Zinc	^{65}Zn	243 days	^{70}Zn	0.62
Molybdenum	^{99}Mo	67 hr	^{94}Mo	9.04

Nuclear technology has been used very successfully in water use efficiency measurements and scheduling of irrigation in irrigated and dry land agriculture. It also helped in search of water resources, ground water hydrology and screening of genotypes of field crops and tree plants efficient for dry land conditions.

In irrigation research, use of neutron moisture meter offered excellent scope in measurement of available moisture status of soil in-situ. This is useful in scheduling irrigation for different crops under different management conditions. There is no other conventional or classical method which can measure so fast and accurately the available moisture status of soil. Neutron moisture meter is a user's friendly equipment which can monitor the moisture status of soil both under irrigated and rainfed management conditions. With the help of this equipment the water use efficiency in dry land agriculture can be substantially increased.

Nuclear Agriculture: Improved seed varieties

	CROP	NO.	CHARACTERISTICS
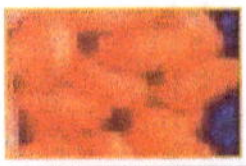	GROUNDNUT	9*	High yielding, improved quality
	PIGEON PEA	2	High yielding, disease resistant, early maturing improved quality
	BLACKGRAM	4	High yielding, disease resistant
	MUNG BEAN	4	High yielding, disease resistant
	RICE	1	High yielding improved quality
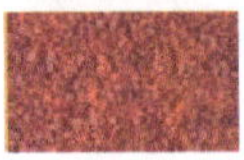	MUSTARD	2	High yielding, improved quality
	JUTE	1	High yielding, fibre yielding

* *Latest updated data*

Crop Varieties Developed at BARC Using Mutation Breeding

10.1.2 Plant Breeding and Genetics

A remarkable progress has been made in the genetic improvement of crops through induced mutation in combination with several other modern biotechnological methods. In several countries including India significant biotechnological improvements have been achieved in rice, wheat, barley, maize and several horticultural crops in terms of crop quality and yield. In genetic improvement of oilseed and industrial crops, induced mutations resulted in improving plant architecture, oil content and oil quality. Several mutant varieties have been released in India, Bangladesh and China and particularly sesame varieties in Egypt and Republic of Korea in recent past. In Africa, mutation breeding resulted in improving yields of local cultivars of Cassava, yam, sweet potato, plantain, sorghum, African rice and several legumes. In India about 76 varieties of different mutant crop cultivars covering cereals, pulses, oilseeds, vegetables, fiber crops, sugar crops, spices etc. have been released or approved for commercial cultivation. All these cultivars have been developed using physical mutagens, gamma rays being the most successful followed by X-rays (Deb, 1997). In Europe most of the promising barley varieties used for making whisky have been developed by inducing mutation in recommended cultivars of barley by physical mutagens. These results are sufficient to indicate that radiation techniques can be employed in improving crop quality with respect to carbohydrate, protein and oil according to human needs in the fields of human nutrition, animal feeding and industrial applications. IAEA has provided direct help and service to its 48 member states for genetic improvement of crop genotypes for sustaining agricultural production.

Nuclear Agriculture Tissue Culture

Pineapple and Banana varieties developed at BARC using tissue culture technique.

Sesbania rodtrata plant for green manure

Special mention may be made about the development of C-13 discrimination technique by IAEA-FAO CRP programmes in selecting crops, crop varieties and tree species which are highly efficient in the use of limited resources of water and give high yield in the arid and semi-arid regions of Africa, especially in Morocco, Tunisia and Sudan. The C-13 discrimination technique has not yet been utilized by our scientists for screening crop varieties for our dryland agriculture. We are confident that our scientists will be able to utilize all the available nuclear and related techniques of crop production to bring a breakthrough in dryland agricultural research to develop a sound economically viable and ecologically sustainable low input agrotechnology for crop production for dryland agriculture which may ultimately lead to user for second green revolution in Indian agriculture. Effect of gamma radiation on two verities of Zinnia was studied in M_1 generation under field conditions by Kole and Meher (2005). They stated that plant height, no. of branches, pedicel length and flower diameter of zinnia increase at lower dose of 5 kR followed by decreased at higher 10 kR onwards as compared to non irradiated control.

10.1.3 Insect and Pest Control

Achievement in the use of radiation and radioisotopes for insect pest management is remarkable. Sterile insect technique is considered as one of the significant achievement of applied entomology for controlling stored grain pest. Tsetse fly, fruit fly and pink bollworm, screw worm, potato tuber moth, etc. Radioisotopes offer a powerful tool in understanding the fate of pesticides in soils and crops. The radioactive isotopes acts as very versatile tool and are being increasingly used in all over the world in the field of insect ecology (Bhattacharyya *et al.*, 2004). Research with radiations and radio isotope have added knowledge to the understanding of insect behavior, physiology, ecology and radiation biology and these knowledge has been employed in developing better control techniques of insect-pest. Radioactive isotopes, when used as a marker for insect-pest, parasitoids and predators have helped in obtaining valuable information on their fight range, dispersal pattern, population build up etc. This information is very vital for rationalizing control options for integrate pest management.

Development of radiation induced F_1 sterility for control of Lepidoptera pests which are damaging major agricultural and commercial crops in tropical and subtropical countries is another remarkable achievement. Prospects for increasing the economic utilization of SIT have been recently enhanced by the successful development of "Vienna 42" genetic sexing strain of medfly. IAEA is providing a tremendous support to developing countries in Africa to eradicate tsetse flies by supplying more than one million sterile tsetse flies in their SIT (Sterile Insect Technology) programmes.

Weevil Resistant Sweet Potato developed by the Central Tuber Crop Research Institute, Thiruvananthapuram

10.1.4 Agrochemicals and Residues

Radioisotopes offer a powerful tool in understanding the fate of pesticides in soil and crop. Monitoring of pesticide levels in food and the environment to support the enforcement of relevant legislation is an essential follow-up measure to 'Agenda-21 of the UN conference on Environment and Development. Use of isotopes has become absolutely essential to study the effects of organochlorine insecticides on agricultural ecosystems of the developing countries. It is heartening to report that the IAEA and FAO are encouraging the development of controlled release and other formulations of pesticides using nuclear techniques. IAEA has developed a formulation of deltamethrin for use on target sources deployed in tsetse fly radiation programmed in Africa. In this formulation deltamethrin retains its insecticidal properties for much longer period under field condition than in commercially available formulations. A similar controlled release formulation has been developed for herbicide thiobencarb for use in paddy fields.

10.1.5 *Food Preservation by Irradiation*

Using gamma irradiation techniques, our scientists have developed new technology for prevention of post harvest losses of food grains, wheat products, pulses, fruits, vegetables, mushrooms, fishes, poultry and meat, frozen seafood and spices. Multi-generation feeding trails with irradiated food on experimental animals have proved the safety of irradiated food. In 1980, FAO/IAEA/WHO Joint Expert Committee on Wholesomeness of irradiated foods declared that all foods irradiated up to 10 kGy is safe for human consumption and need no toxicological evaluation. This declaration stimulated further research and developmental activities in food preservation using irradiation techniques. In August 1994, Ministry of Health, Government of India through a Gazette Notification, permitted irradiation processing of onions, potatoes and. spices in India for domestic consumption.

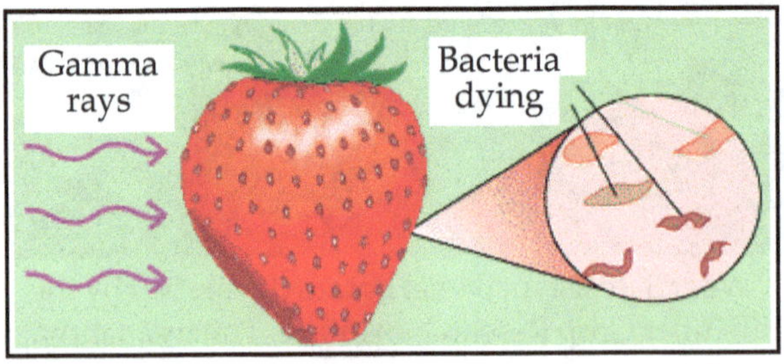

The IAEA and FAO continued their efforts to facilitate the large scale use of food irradiation on a commercial scale in developing countries where storage losses are increasing at an alarming rate and death due to food poisoning, parasitic and bacterial diseases are most frequent. Irradiation of fruits and vegetables by gamma rays has economic advantages for disinfestations over the use of methyl bromide the most widely used fumigant for food and agricultural produce. Although the world of scientific community has declared the safety of irradiated food up to 10 kGy dose of irradiation, the consumer acceptance of the irradiated food is not of very high order in developing countries. Participants from 13 countries in a CRP reported successful marketing of several irradiated food items and reported that irradiation has been used for processing several food items for commercial purpose in China, Indonesia, Japan, the Republic of Korea and Thailand. IAEA has built large demonstration irradiation facilities in Bangladesh, China and Vietnam and one commercial irradiator has been built in Indonesia. According to IAEA, more than 1, 30,000 tonnes of food have been irradiated in these countries during the past five years. In India, BARC has initiated steps to commercialize food irradiation on a large scale. Irradiation preservation of cereals, pulses, fruits, vegetables and many other food materials for domestic consumption is under active consideration of Govt. of India and a Government notification to this effect is expected in near future. Irradiation of mango fruits with 200 Gy gamma radiations increased the post harvest life of 93.75 per cent mango fruits by 12 days over control (46.66 per cent) and proved superior to maintain the considerable physico-chemical composition of mango fruits as compared control and higher dose of gamma radiation. (Baghel *et al.,* 2005).

Radiation Technology for Food Preservation

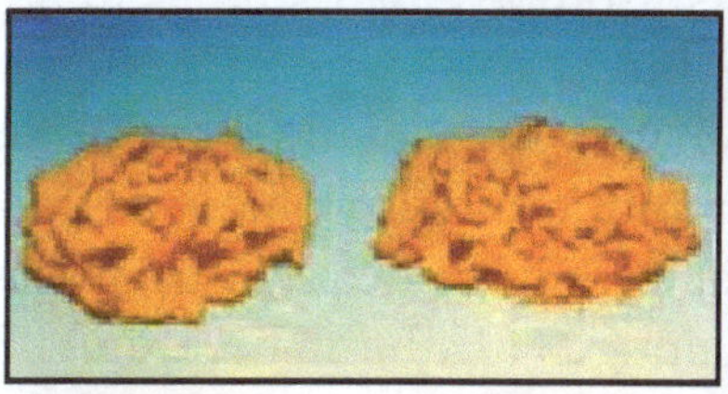

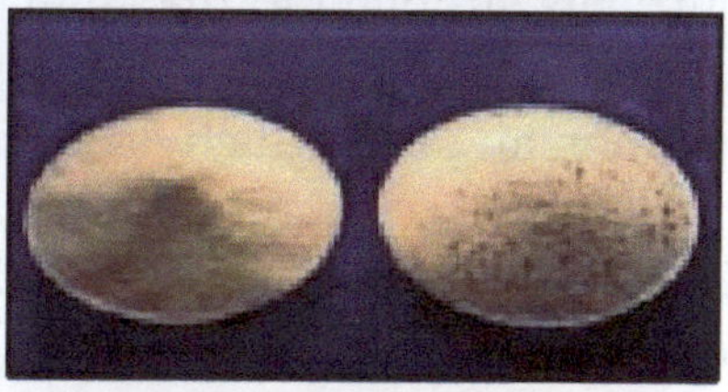

Radiation processing has been established as a simple and very effective technique for preservation of various food items. Shown on the left in the pictures are radiation processed potatoes, onions, turmeric, sooji and banana.

Some of the food items which have been approved by the Government of India for radiation processing.

Food irradiation applications

Low dose (up to 1 kGy)	Inhibition of sprouting	Potatoes, onions, garlic, ginger, yam
	Insect and parasite disinfestations	Cereals, fresh fruit, dried foods
	Delay ripening	Fresh fruit, vegetables
Medium dose (1-10 kGy)	Extend shelf life	Fish, strawberries, mushrooms
	Halt spoilage, kill pathogens	Seafood, poultry, meat
High dose (10-50 kGy)	Industrial sterilization	Meat, poultry, seafood, prepared foods
	Decontamination	Spices, etc

10.1.6 Animal Production and Health

The animal production mainly depends on the quality and quantity of nutrients supplied to the animals. Isotope aided studies in animal nutrition have led to spectacular increases in growth rates, milk production and reproductive efficiency by improving the nutrition of ruminant livestock through incorporation of supplementary feeding with locally available agro-industrial products such as urea, molasses, oil cakes, legume tree leaves etc. along with the main animal feed. Since individuals within an animal population have different nutritional requirements, isotopic methods can be planned for individual animals to define their nutritional and reproductive problems with special reference to production of milk, meat and fiber and to find solution of problems in such a way that the cost benefit ratio of the evolved technologies are favorable to the poor farmers for adoption.

Isotope aided immunoassay test kits have been developed successfully for diagnosing and surveillance of selected viral, bacterial and parasitic diseases of livestock in developing countries. EIISA diagnostic kits have been supplied by IAEA to detect antibodies in cattle blood for seromonitoring of rinderpest disease in several African countries. Mani and Chandra (2005) reported that feeding irradiated soybean at 20 kGy may reduce the NH_3-N concentration which may be beneficial for better utilization of diet by goats.

10.1.7 Nuclear techniques in rapid analysis of soil, plant and food

Nuclear techniques are also being used in Indian for rapid analysis of soil, plant and food materials with high degree of accuracy. Special mention may be made of the rapid non destructive method of analysis of oil content in oilseeds by pulsed NMR technique without weighing and drying the seeds. The NMR technique is now being used for screening drought tolerant varieties of wheat.

The achievements of our scientists in the peaceful uses of nuclear and related techniques in Indian agriculture is remarkable and it gives us a great hope that in the coming years nuclear and allied techniques will play major role in sustaining Indian agriculture and environment. Radiation technology and Isotope techniques have major application practically in all areas covered by Agenda 21 of the Earth Summit in Rio in June, 1992 and "will continue to be put to good use for the benefit of humanity, both in material terms and for improving the quality of life".

10.2 Radioisotopes in Chemical Biology

10.2.1 Metabolism activity

(1) Metabolic pathways

Radioisotopes are frequently used for tracing metabolic pathways. This usually involves adding a radioactive substrate, taking samples of the experimental material at various times, extracting and chromatographically, or otherwise, separating the products. By identifying the labelled compounds containing the radioactivity in each of them and plotting suitable graphs, it is possible to obtain considerable information on the pathway involved.

(2) In photosynthesis

With the discovery of C-14 by Ruben and Kamen in 1940 in California, attempts were made to understand how CO_2 and H_2O are converted into complex energy containing materials by green plants in the presence of sunlight. Calvin in 1946 studied photosynthesis by algae (chlorella) in a nutrient medium containing C-14 labelled bicarbonate solution. At different periods of photosynthesis varying from few sec to 30 sec, the compounds synthesized by the plant were isolated by paper chromatography and autoradiographs were taken. At 5 sec exposure, 6 stable compounds were formed. The first stable compound formed is PGA (3-phosphoglyceric acid) in all C_3 plants.

Path of carbon 14 in photosynthesis

The work with C-I4 further indicates that some amino acids were formed in the course of photosynthesis since these often appear before the carbohydrates. The radioautograph revealed the presence of amino acids such as alanine and aspartic acid with 10 sec exposure time. There are some evidences, too, that fats aid formed in the photosynthetic cycle. After Calvin- Benson or C_3 pathway, another pathway was discovered using C-14 tracer which is known as Hatch-Slack or C_4 pathway. In this process the CO is fixed initially into 3C PEP catalyzed by the enzyme PEP carboxylase to form oxaloacetate which is finally converted into malic acid and aspartic acid. 'The C_4 plants are most common in warm environments where water may be limiting, e.g. com, sugarcane, sorghum and many grasses.

In identifying the C_3 and C_4 plants, use of C-13 isotope has also shown tremendous potentiality.

(3) In lysine biosynthesis

That a functional DAP (2,G-diaminopimelate)pathway for lysine biosynthesis is operative in plants was elegantly illustrated by using ^{14}C labelled precursors and following the labelling in the different carbons of the lysine bio synthesized. Earlier, it was believed that a lysine pathway identical with the one known in fungi was operative in higher plants. This was because; one of the key intermediates, AAA (a-aminoadipii acid) was isolated from several higher plants. By using the tracer it becomes obvious that AAA was formed as a result of catabolism and not anabolism.

(4) In metabolic turnover times

Radioisotopes provide a convenient method of ascertaining turnover times for particular compounds. For example, a group of rats was injected with a radioactive amino acid and left for 24 hours, during which time most of the amino acid will be assimilated into proteins. The rats are then sacrificed at suitable time intervals and radioactivity in organs or tissues of interest is determined. In this way it is possible to ascertain the rate of metabolic turnover of protein. Using this sort of method it has been shown that liver protein is turned over in 7 to 14 days, while skin and muscle protein is turned over every 8 to 12 weeks and collagen is turned over at a rate of less than 10% per annum.

10.2.2 Absorption, accumulation and translocation

Radioisotopes have been very widely used in studying the mechanisms and rates of absorption, accumulation, and translocation of inorganic and organic compounds by both plants and animals. Such experiments are generally simple to perform and can also yield evidence on the route of translocation and sites of molecules accumulation of biological interest.

10.2.3 Analytical applications

Enzyme and ligand binding studies

Virtually any enzyme reaction can be assayed using radiotracer methods, provided that a radioactive form of the substrate is available. Radiotracer based enzyme assays are more expensive than other methods, but frequently have the advantages of a higher degree of sensitivity. For example, activity of aspartokinase can be studied by using ^{14}C- aspartate as the substrate. This is converted, in vitro to β aspartyl hydroxamate-^{14}C in the presence of hydroxylamine. The radioactive β- aspartyl hydroxamate can be separated from the substrate by paper or thin layer chromatography.

10.2.4 Isotope dilution analysis in closed system

There are many compounds present in living organisms which cannot be accurately assayed by conventional means, since they are present in such low amounts and in mixture of similar compounds. Isotope dilution analysis offers a convenient and accurate way of overcoming this problem and avoids the necessity of quantitative isolation. For instance, if the amount of iron in a protein preparation is to be determined, this may be difficult by using normal methods, but it can be done if a source of ^{59}Fe is available. This is mixed with the protein and a sample of iron is subsequently isolated, assayed for total iron and the radioactivity determined.

10.2.5 Radioimmunoassay (RIA)

One of the most significant advances in biochemical techniques in recent years has been the development of radioimmunoassay. It combines the specificity of the immune reaction with the sensitivity of radioisotope techniques, and is used for the quantitative analysis of hormones, steroids and drugs.

The technique is based on the competition between unlabelled antigen and a finite amount of the corresponding radio-labelled antigen for a limited number of antibody building sites in a fixed amount of antiserum. At equilibrium in antigen excess, there will be both free antigen and antigen bound to the antibody. Under standard conditions, the amount of labelled antigen bound to antibody will decrease as the amount of unlabelled antigen in this sample increases.

By using known amounts of unlabelled antigen and a fixed amount of antibody and labelled antigen, the amount of labelled antigen bound as a function of the total antigen added is measured and a calibration curve constructed. This calibration curve may then be used to determine the amount of antigen in samples treated similarly.

10.3 Radioisotopes in Molecular Biology

Molecular biology and genetic engineering have recently experienced a revolutionary change with the development of techniques for gene identification, cloning and manipulation. Many of these techniques depend on the hybridization of a nucleic acid probe to a target DNA or RNA sequence. For the detection of the hybrid it is imperative that it be labelled. There are five basic methods for labeling nucleic acids:

1. Nick translation
2. Primer extension
3. Methods based on RNA polymerize
4. End-labelling
5. Direct labelling

10.4 Radioisotopes in Medicine

Radioisotopes are extensively used in nuclear medicine to allow physicians to explore bodily structures and functions *in vivo* (in the living body) with a minimum of invasion to the patient. Radioisotopes are also used in radiotherapy (radiation therapy) to treat some cancers and other medical conditions that require destruction of harmful cells.

Radioisotopes, containing unstable combinations of protons and neutrons, are created by neutron activation involving the capture of a neutron by the nucleus of an atom resulting in an excess of neutrons (neutron rich). Proton rich radioisotopes are manufactured in cyclotrons. During radioactive decay, the nucleus of a radioisotope seeks energetic stability by emitting particles (alpha, beta or positron) and photons (including gamma rays).

Although nuclear medicine traces its clinical origins to the 1930s, the invention of the gamma scintillation camera by American engineer Hal Anger in the 1950s, however, brought major advances in nuclear medical imaging and rapidly elevated the use of radioisotopes in medicine. Radioisotopes allow high quality imaging of bones, soft organs (e.g., thyroid, heart, liver, etc.). A number of diagnostic techniques in nuclear medicine use gamma ray emitting tracers. The tracers are formed from the bonding of short-lived radioisotopes with chemical compounds that allow the targeting of specific body regions or physiologic processes. Emitted gamma rays (photons) can be detected by gamma cameras and computer enhancement of the resulting images allows quick and relatively non-invasive (compared to surgery) assessments of trauma or physiological impairments.

Technetium-99 (an isotope of the artificially-produced element technetium) is a radioisotope widely used in nuclear medical procedures. Technetium-99 decays by an isomeric process which emits gamma rays and low energy beta particles (electrons). Technetium is supplied to hospitals from nuclear reactors in containment vessels initially containing molybdenum-99 that, with a half-life of 66 hours, decays to technetium-99 which is removed by flushing.

American chemist Peter Alfred Wolf's (1923–1998) work with radioisotope utilizing positron emission tomography (PET) led to the clinical diagnostic use of the PET scan. Positron emission tomography utilizes isotopes produced in a cyclotron. Positron-emitting radionuclides are injected and allowed to accumulate in the target tissue or organ. As the radionuclide decays it emits a positron that collides with nearby electrons to result in the emission of two identifiable gamma photons. PET scans use rings of detectors that surround the patient to track the movements and concentrations of radioactive tracers. PET scans have attracted the interest of physicians because of their potential use in research into metabolic changes associated with mental diseases such as schizophrenia and depression. PET scans are used in the diagnosis and characterizations of certain cancers and heart disease, as well as clinical studies of the brain.

Cancer and other rapidly dividing cells are usually sensitive to damage by radiation. Accordingly, some cancerous growths can be restricted or eliminated by radioisotope irradiation. The most common forms of external radiation therapy use gamma and x rays. During the last half of the twentieth century the radioisotope cobalt-60 was frequently used source of radiation used in such treatments. More modern methods of irradiation include the production of x rays from linear accelerators.

Internal radiotherapy involves the introduction of a radioisotope as a radiation source. Iridium-192 implants emit both gamma and beta rays that destroy surrounding target tissue. In low dose forms, strontium-89 has been used to relieve cancer-induced bone pain. Not all radioisotope techniques involve restricted sites, in the treatment of some diseases requiring bone marrow transplants the malfunctioning marrow is killed with a massive dose of radiation before the introduction of healthy marrow.

Because they can be detected in low doses, radioisotopes can also be used in sophisticated and delicate biochemical assays or analysis. There are many common laboratory tests utilizing radioisotopes to analyze blood, urine and hormones. Radioisotopes are also finding increasing use in the labelling, identification and study of immunological cells. Shielded radioisotopes are also used to power heart pacemakers and sterilize medical instruments.

Iodine-131, phosphorus-32 are commonly used in radiotherapy. More radical uses of radioisotopes include the use of Boron-10 to specifically attack tumor cells. Boron-10 concentrates in tumor cells and is then subjected to neutron beams that result in highly energetic alpha particles that are lethal to the tumor tissue.

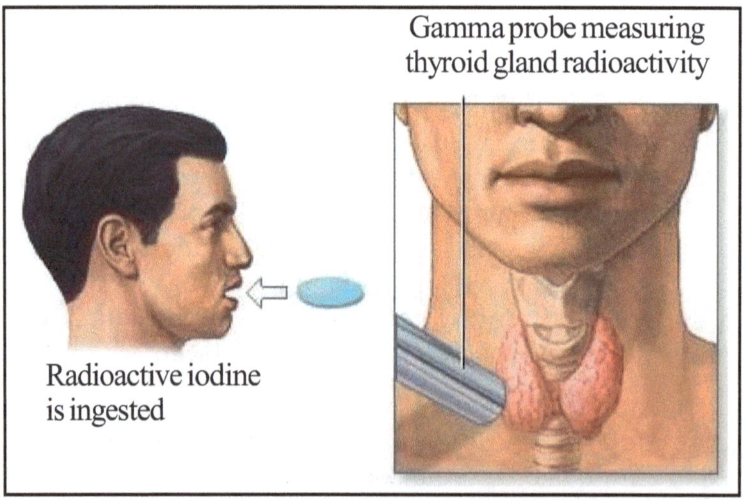

10.5: Radioisotopes in Microbiology and Immunology

The selection of radioisotopes for medical use is governed by several important considerations involving dosage and half-life. Radioisotopes must be administered in sufficient dosages so that emitted radiation is present in sufficient quantity to be measured. Ideally the radioisotope has a short enough half-life that, at the delivered dosage, there is insignificant residual radiation following the desired length of exposure. Regardless, the use of radioisotopes allows increasingly accurate and early diagnosis of serious pathology (e.g. tumors) and earlier diagnosis often results in more favorable outcome for patients.

Radioisotopes, containing unstable combinations of protons and neutrons, are created by neutron activation that involves the capture of a neutron by the *nucleus* of an atom. Such a capture results in an excess of neutrons (neutron rich). Proton rich radioisotopes are manufactured in cyclotrons. During radioactive decay, the nucleus of a radioisotope seeks energetic stability by emitting particles (alpha, beta or positron) and photons (including gamma rays). The history of radioisotopes in microbiology and *immunology* dates back to their first use in medicine. Although nuclear medicine traces its clinical origins to the 1930s, the invention of the gamma scintillation camera by American engineer Hal Anger in the 1950s brought major advances in nuclear medical imaging and rapidly elevated the use of radioisotopes in medicine.

The study of *microorganisms* also relies heavily on the use of radioisotopes. The identification of protein species, labeling of surface components of *bacteria,* and tracing the *transcription* and *translation* steps involved in nucleic acid and protein manufacture all utilize radioisotopes.

The various types of radiations can be selected to provide information on a sample of interest. For example, to examine how quickly a protein is degraded, an isotope that decays very quickly is preferred. However, to study the adherence of bacteria to a surface, a radiolabel that persisted longer would be more advantageous.

Furthermore, various radioactive compounds are used in microbiological analyses to label different constituents of the bacterial cell. Radioactive hydrogen (i.e., tritium) can be used to produce radioactive *DNA (deoxyribonucleic acid).* The radioactive DNA can be detected by storing the DNA sample in contact with x-ray film. The radioactive particles that are emitted from the sample will expose the film. When the film is developed, the result is an image of the DNA.

This process, which is known as autoradiography, has long been used to trace the elongation of DNA, and so determine the speed at which the DNA is replicating.

DNA can also be labeled, but in a different location within the molecule, by the use of radioactive phosphorus.

Bacterial and viral proteins can be labeled by the addition of radioactive methionine to the growth mixture. The methionine, which is an amino acid, will be incorporated into proteins that are made by several paths can then be followed. For instance, in what is known as a pulse-chase experiment, the radioactive label is then followed by the addition of nonradioactive (or "cold") methionine. The rate at which the radioactivity disappears can be used to calculate the rate of turnover of the particular protein. In another experimental approach, the protein constituents of bacteria or viruses can be separated on an electrophoretic gel. The gel is then brought into contact with x-ray film. Wherever a radioactive protein band is present in the gel, the overlaying film will be exposed. Thus, the proteins that are radioactive can be determined.

The use of radiolabelled compounds that can be utilized as nutrients by bacteria allows various metabolic pathways to be determined. For example, glucose can be radiolabelled and its fate followed by various techniques, including chromatography, autoradiography, and gel electrophoresis. Furthermore, a molecule such as glucose can be radiolabelled at various chemical groups within the molecule. This allows an investigator to assess whether different regions of a molecule are used preferentially.

Radiolabelling has allowed for great advances in microbiological research. A well-known example is the 1952 experiment by Hershey and Chase, which established that DNA was the reservoir of genetic information. Bacterial viruses were exposed to either radioactive sulfur or phosphorus. The sulfur radiolabelled the surface of the virus, while the phosphorus labeled the DNA. Viruses were allowed to infect bacteria and then were mechanically sheared off of the bacteria. The sheared viruses were then collected separately from the bacteria. Radioactive sulfur was found in the virus suspension and radioactive phosphorus was found in the bacteria. Furthermore, the bacteria eventually produced new virus, some of which had radioactive DNA. Thus, radio labelling demonstrated the relationship between DNA and genetic information.

10.6 Radioisotope in Marine Science

A distinct advantage of the mouse bioassay is that it provides an integrated measure of total toxicity in a sample. This is an issue because each of the toxin poisoning syndromes results from the combined actions of distinct families of toxins. PSP (Paralytic Shellfish Poisoning), for example, results from a family of compounds called the axitoxins, consisting of more than 20 related compounds or derivatives, each with a different intrinsic potency that vary by over three orders of magnitude. Analytical instruments such as HPLC can separate and quantify these individual compounds, but do not provide a simple and rapid measure of toxic potency, as does the bioassay. Likewise, antibody-based assays do not provide an accurate estimate of toxicity because the affinity of those antibodies for the multiple toxin derivatives in a sample varies with toxin structure in a way that may not reflect potency.

A key application of nuclear technologies that can circumvent the problems highlighted in the preceding paragraphs is the receptor binding assay or RBA. The detection of HAB toxins using an RBA is based on the function, or pharmacological activity of the toxins ñ i.e., the highly specific interaction of a toxin with its biological receptor. Like the animal bioassay, the receptor-binding assay provides an estimate of the integrated toxic potency of a sample. Other advantages of the receptor assay include limited equipment costs and ability to work with crude (unpurified) extracts.

Radioisotope Methods for Studying Contaminant

A) Biogeochemistry

To varying degrees, edible marine organisms concentrate inorganic and organic contaminants from water, their food or sediments. Understanding the transfer of contaminants through the food web is critical to predicting the exposure of humans to contaminants (either through subsistence or commercial consumption of seafood) and the possible health consequences of such exposure. In addition, such information is crucial to making accurate risk assessments for seafood safety purposes, a topic which is attracting much attention. Risk assessments based on sound science are the key to cost-effective management strategies.

The use of radiotracers to obtain such information involves exposing organisms to radiolabelled forms of the contaminant of interest, often trace metals. The amount of tracer incorporated into the organism is followed over time using gamma spectrometry. Bio-accumulation potential is rapidly screened, and assimilation or uptake efficiencies can be calculated. Likewise the depuration rates and half-times of the radio-labeled contaminant in the live organism can be followed after exposure. Exposing organisms to radiolabelled food such as phytoplankton, detritus or larger food matter and following the distribution and fate of the radiotracer as it moves through a food web can elucidate trophic transfer pathways. In some cases, use of selected stable isotopes of the heavy metal in question, measured for example by ICP-MS, can also give important insights into contaminant behavior and fate; however, the chemical analysis involved requires destruction of the samples. Using a similar experimental design, radio-labeled organic contaminants (e.g. sticides), most often tagged with^{14}C, can be monitored over time by liquid scintillation counting techniques although this radio-analytical methodology also is necessarily destructive and therefore time-consuming. The use of radiotracers to determine uptake of heavy metal contaminants by organisms under controlled laboratory conditions is a rapid and sensitive, non-destructive approach. Moreover it offers the opportunity to determine uptake patterns and rates in living organisms using environmentally realistic concentrations of the contaminant of interest. Although the radiotracer approach is not unique, it is preferred in many applications. The property of radioactivity offers the ability to detect low concentrations of the contaminant in water, sediment or organism tissue. Some disadvantages include the relatively high cost of the radiotracers and the fact that the radioactivities are significant, requiring their regulation and the establishment of safety protocols for laboratory handling. The technique also requires the maintenance of organisms in a controlled laboratory setting

The major alternative determining uptake rates with the stable form of a metal requires using elevated mass concentrations to improve detection, clean laboratory conditions to avoid contamination, acid digestions, and exhaustive wet chemical analyses.

Although as noted above, alternatives exist to using nuclear techniques to determine rates of contaminant uptake and retention, the nuclear approach is rapid and mature. It is transferable to developing countries and sustainable (although expensive) once transferred. The approach complements other methods of assessing contaminant uptake and lends itself especially well to providing information on parameters needed to model uptake and trophic transfers of contaminants. The MEL has a dedicated in-house capability in this approach and thus can provide significant advice and guidance in transferring it to a developing country. The approach can thus be considered as having high applicability but requires considerable Government investment on an ongoing basis.

B) Ecosystem Health

Nuclear technology can contribute in various ways to assessment of the health of an ecosystem. But in the context of technology transfer, the technology's applicability varies widely depending on the particular application. Nuclear techniques are particularly well suited for studying several important aspects of habitat modification and loss. For example, the input and accumulation of sediments and associated contaminants to a coastal area can have impacts on habitat. Natural and anthropogenic radionuclides that associate with particles may be used to establish sediment chronologies for reconstructing the history of contaminant inputs. It is also possible to resolve atmospheric and non-atmospheric sources of contaminants through a comparison of contaminant and radionuclide (^{210}Pb) inventories in sub-total and wetland sediments.

The nuclear techniques are mature and well established for such applications. Indeed, their "uniqueness" is high because there are few alternatives to nuclear techniques for determining high-resolution chronologies of coastal sediments. Once the capability is established in a developing country, there are numerous applications of these approaches that make them complementary to other applications. Training is necessary to avoid pitfalls in determining chronologies (for example due to the mixing effects of in fauna), but such training courses already exist. Thus this approach has high applicability.

Nuclear techniques also lend themselves to assessing the extent to which wetlands are responding to changes in sea level. Global increases in the rate of sea level rise have caused concomitant increases in the accretion rate of coastal wetlands such as salt marshes. The application of radionuclides to determining accretion rates of wetland deposits is analogous to the approach used for sub-tidal sediments. The technique applied to wetlands has a comparable strategic value and thus has high applicability. Finally, changes in circulation and current patterns can affect the distribution of passive marine biota such as phytoplankton and certain larval forms. Numerous isotopic techniques exist to trace circulation. These range from the use of naturally occurring radionuclides such as the Ra isotopes to anthropogenic ^{137}Cs to stable isotopes (e.g. of O), as well as radiotracers added purposefully. Many alternatives exist to the use of these nuclear techniques and perhaps their greatest utility is in validating models of circulation and dispersal. As a consequence, the techniques do not have a high priority for technology transfer.

C) Eutrophication/Hypoxia and HABs

Eutrophication is an important coastal zone problem, as it often leads to degraded water quality, including hypoxia and harmful algal blooms, with consequent impacts on living resource. Assessment of the sources of nutrients and organic matter in a coastal ecosystem is critical to understanding the causes of eutrophication. Two aspects of eutrophication are especially suitable for the application of nuclear techniques; these are the determination of the sources of nutrients and of organic matter.

D) Changes in the abundance of exploitable living marine resources

The abundance of living marine resources is affected by many factors, some of which may be addressed using nuclear approaches. In particular, changes in the frequency and intensity of coastal upwelling can directly affect fishery production and yield. Such changes can be assessed through a combination of radiometric sediment core chronologies and stable C isotopes or biomarkers that gauge marine organic matter production. The use of sediments to conduct retrospective studies of up-welling history and associated primary productivity changes might yield valuable insight into the functioning of individual systems and their fisheries. Comparisons of similar databases between different up-welling systems should permit the verification of connections brought on by such events as El Niño. Such comparisons would aid our understanding of the effects climate change has on the functioning of the oceans and their living resources. Nuclear approaches have a high applicability in this area because they are uniquely well suited to determining sediment chronologies.

Changes in the structure of marine food webs also can affect the abundance of living marine resources. Information on trophic relationships is very important for coastal planners and managers. Indeed, this is essential when considering comparative productivity of discrete localities; biological diversity and intrinsic ecological values; inter-dependencies of particularly important species (whether for commercial, subsistence or ecological value); and potential for bioaccumulation of contaminants. Among the tools most commonly used for such studies are various sampling techniques and fish-gut analyses. The use of stable isotopes (C and N) can sometimes reveal subtleties such as "short circuits" in the food chain that may not be discernible with the more common methodologies. Nuclear/isotopic techniques are also capable of faster results. Nonetheless, nuclear/isotopic techniques are not notably cost-effective when applied to the study of trophic relationships, and the results tend to be site specific.

E) Changes in landings

Knowledge of fish population dynamics is key to understanding changes in landings. Indeed, management of commercial fisheries is highly dependent upon life history investigations, particularly studies of populations' age structure. "Reading" the number of annuli on scales, otoliths or other hard body parts are a typical method for determining age. However, such traditional techniques can sometimes prove totally ineffective, and therefore need verification. This is especially true in the tropics, where water temperature is less variable, thus providing conditions less conducive to the formation of annuli. Even some cold water or deep-water species may exhibit similar problems.

Naturally occurring radioisotopic tracers are an effective means of determining the ages of fish, but the technique is not widely used. Isotope technology for conducting such studies is both cost effective and technically mature, and effectively complements traditional methodologies. It is thus judged as highly applicable and appropriate for technology transfer.

F) Effects of mariculture

Mariculture brings new challenges to the coastal zone. One of the major impacts of fish farms, and of mariculture in general, is the localized introduction to the ecosystem of highly labile organic matter in the form of uneaten fish food, ammonia (animal excretion products), feces (fishes, bivalves) and pseudo-feces (bivalves). This continuous organic input can eventually lead to eutrophication of the production site. The continuous input of organic matter gradually changes the biogeochemical conditions of the seafloor. These changes are initially undetected in the water column characteristics. However, the final state is anoxia accompanied by liberation of poisonous gases such as H_2S and the death of native and cultured species. Such biogeochemical changes can be studied by using naturally occurring radionuclides to determine the rates of bio-irrigation and biological mixing in bottom sediments. These two rates determinations can then be combined with measurements of sediment oxygen profiles. All three measurements together provide a better biogeochemical characterisation of the sea floor in response to incoming organic matter than any one of them alone. Another important aspect that can affect the future of a mariculture activity is a well- characterized production site. Traditionally, oceanographic tools (currents, nutrients, bathymetry, etc.) have been extensively applied to answer this need. However, radiological techniques are a highly desirable complement to traditional methods. Radiochronology and sediment accumulation rates are powerful tools for characterizing past input fluxes from both natural and anthropogenic sources.

Nuclear techniques provide tools that may help emerging and established seafood farmers to comply with national and international regulations governing the use and production of farmed fish and shellfish. Nuclear approaches are highly applicable to address such concerns, but have not yet been widely applied for this specific purpose.

10.7 Radioisotopes in Hydrology

Lawrence Livermore scientists use both stable and radioactive isotopes to learn about groundwater sources, ages, travel times, and flow paths and to determine the path and extent of contaminant movement in the water. The combination of Livermore's unique nuclear chemistry experience, multidisciplinary staff, and highly sensitive mass spectrometry equipment, including use of the Center for Accelerator Mass Spectrometry, means that Livermore can take the measurements and interpret the results effectively. Livermore is unique among isotope hydrology centers in having all of the various tracing methods in one place. Researchers routinely use a variety of isotopes to evaluate a groundwater system and create the most accurate picture possible.

10.7.1 The Water's Origin

All waters have "fingerprints" of naturally occurring isotopes that provide information about their origin. Among the most powerful and cost-effective fingerprinting tools are the ratios of stable isotopes of hydrogen—deuterium to hydrogen (D/H)—and of oxygen-18 to oxygen-16 ($^{18}O/^{16}O$). For example, the D/H and $^{18}O/^{16}O$ ratios in precipitation vary according to elevation and distance from the ocean. An altitude difference of 250 meters (820 feet) produces a clear and measurable change in the two ratios, which is preserved once the precipitation infiltrates to the aquifer. While some mixing of waters in the aquifer is inevitable, an aquifer is typically not homogeneous. Layers of heavy clay retard the movement of groundwater within the aquifer and often serve to keep waters from different sources distinct from one another. Thus, scientists can use D/H and $^{18}O/^{16}O$ ratios to determine the recharge location for a groundwater and to discriminate among multiple water sources within an aquifer. The concentrations of atmospherically derived noble gases (neon, argon, krypton, and xenon) in groundwater provide another excellent fingerprinting tracer because the noble gases do not react with surrounding materials and because their concentrations are preserved as they are recharged into the aquifer. The relative abundances of the noble gases provide information about the temperature and altitude at the time of recharge. Isotopically enriched noble gases, which are commercially available at low cost, can also, be used as nonhazardous artificial tracers. They are detectable at very low concentrations, rendering them highly effective for delineating groundwater flow systems.

10.7.2 The Water's Age

Naturally occurring radioactive isotopes provide information about a groundwater's age, which refers to the last time the water, was in contact with the atmosphere. Once the recharge area has been established using isotope fingerprinting methods, radioisotopes may be used to estimate how long it took for a parcel of groundwater to travel from its recharge area to the measurement point. For young groundwater, the best dating method is one that measures the relative abundance of tritium (^{3}H) and helium-3 (^{3}He). Tritium is a radioactive isotope of hydrogen that decays to 3He with a half-life of 12.4 years. The amount of ^{3}He from the decay of tritium is measured along with the amount of tritium remaining in the water. That sum is equal to the amount of tritium that was present at the time of recharge. The ^{3}H-^{3}He dating method is remarkably accurate for ground waters up to 40 years old with an age resolution of plus or minus one year.

The ages of older ground waters can be measured in several ways. One is by measuring the carbon-14 (^{14}C) dissolved in the water. Carbon-14 decays at a known rate with a half-life of 5,730 years, providing a useful dating method for ground waters less than 40,000 years old. Another method uses the isotope helium-4 (^{4}He), which is produced continuously deep within the earth by the radioactive decay of uranium and thorium. Relative age relationships determined from the amount of ^{4}He in a groundwater are a valuable supplement to ^{14}C age determinations. The long-lived radionuclides chlorine-36 and iodine-129 are also being used to study very old regional groundwater flow systems.

10.7.3 Rock and Contaminants

Other isotopic and elemental measurements provide useful information about the interaction of water and its host rock or sediment or about the migration of contaminants in the water. For example, the isotopic composition of strontium (Sr) in groundwater depends on the $^{87}Sr/^{86}Sr$ ratios of the rocks with which it interacted. (Different rock types can have significantly different $^{87}Sr/^{86}Sr$ ratios.) In basins containing several aquifers with differing rock characteristics, the dissolved $^{87}Sr/^{86}Sr$ of a water sample may help distinguish where a particular groundwater resides.

Pollutants in groundwater can be studied in several ways. Measurements of nitrogen and oxygen isotope ratios ($^{15}N/^{14}N$ and $^{18}O/^{16}O$) can supply information about the source of nitrates, which are prevalent in agricultural fertilizers and may contaminate groundwater. Similar information about manmade pollutants can be gathered by measuring lead isotopes and heavy metal concentrations.

❑❑❑

Chapter 11

Laboratory Manual

In the Radiochemical Laboratory

The following are some illustrative recommendation which may be used selected as a guide for setting up working rules :

General

- No unnecessary materials should be brought into the laboratory. No unnecessary work should be done there.
- Eating, drinking, and the use of cosmetics in the laboratory should be forbidden. Radioactive materials should not be place in milk bottles, soft drink bottles, or other vessels that might be inadvertently used for food purposes.
- Record should be kept of radioactivity brought into the building and of its disposition.

External Hazards

- Radioactivity should be stored so that the general background is less than of the order of 1 mr/hr.
- Work should be planned so as to minimize exposure to external irradiation. In no case should the hands receive more than 1r/ week, nor the whole body more than 300 mr/week. Packed ionization chambers and survey meters should be employed to estimate the dosages received, and thee values should be recorded.

Contamination and Internal Hazards

- Laboratory protective clothing (laboratory coats, gloves, goggles and shore covers) should be employed as need, but these items should not be taken outside the laboratory and worm in clerical offices, rest rooms counting room etc. No person who has breaks in the skin on his hands should work rubbers gloves. Gloves should always be handled so as to avoid contamination of the inside surfaces.
- Operations involving radioactive materials higher than counting levels should be performed over trays bedded down with disposable absorbent paper (diaper paper). Radioactive solutions should be handled in double containers so as to avoid spread of contamination in case of breakage.
- Pipetting or the performance of any similar operation by mouth suction should be prohibited. Glass blowing in the laboratory should be discouraged.
- Hoods should be used when radioactive solutions are to be distilled or heated or when there is formation of volatile compounds or spray. Hood fans should not be turned off while the hood contains radioactive materials.
- When dusty radioactive materials are handled, a dry box should be employed, or if the operation is performed in a hood, the worker should wear a face mask.
- If, in the course of the work, personal or laboratory contamination is suspected, a survey with a suitable instrument should be made immediately; to be followed by required cleaning. Group I and II radioisotopes should be reduced to a level of less than 1000 counts/min, and group III to less than 100 counts/min, as measured with a Geiger counter of about 2 - in^2 flat window area.
- In the event of a spill, the liquid should be blotted up with absorbent paper, when the surface is dry, a survey should be made with an appropriate survey mater and the surface then cleaned until the permissible level is reached.

Disposal

- Solid waste and contaminated materials should be place in properly labeled Trash cans for disposal according to regulations.

- Radioactivity should not be put into the chain or contaminated apparatus washed in the sink unless the levels have been calculated as appropriate for disposal by this route. Rubber tubing attached to the faucet should not be allowed to dip below the surface of the waste in the sink and thus allow the waste to be siphoned back into water lines in the event of water-pressure failure.
- Contaminated glassware or equipment should not be returned to central stores for general use or sent away from the laboratory for repair.

Upon Leaving Laboratory

- Open radiation sources should be plainly marked for the protection of custodial or repair men.
- Before the laboratory is left, the hands should be washed and checked with a suitable low level scanning instrument.

In the Chemical Laboratory

It is perhaps not out of place to list a few safety precautions which apply to all chemical laboratory operations but which may become increasing importance when radioactive materials are involved

Every in the laboratory should familiarize himself with the types, locations and use of fire extinguishers. Sand should be available for putting out metal or metal hydride fires.

1. When flammable solvents are in use, smoking and open flames are forbidden. All electrical equipment should be spark proof of explosion proof. A warning sign should be posted in the doorway.
2. Flammable or highly volatile solvents should not be stored in any laboratory room in greater than 10-gal. quantities, with no more than 5 gal. to a container. (This refers to the total quantity of all solvents in the room).
3. Large quantities of flammable waster liquids that are not miscible with water should be collected metal cane.
4. The contact of boiling, undiluted perchloric acid or perchloric acid with organic matter or easily oxidized inorganic matter results in serious explosions. Consequently such materials must be destroyed before evaporation with perchloric acid is carried

out. Evaporation should always be done in a hood with a good draft. Perchloric acid should never be fumed in hoods made of wood or similar organic material. The quantity of perchloric acid stored in the laboratory should be kept to an absolute minimum and should be stored in such a manner that there is no possibility of contact with organic materials.

5. Concentrated acid and alkalies should be stored below munch-top level in bottles set in lead or enamelware trays.
6. Cleaning solution should be kept covered at all times. Location near the sink should be avoided if possible.
7. An intensive effort should be made to avoid mercury spills and to keep mercury from collecting in various corners of the table.
8. Both liquid oxygen and liquid nitrogen are often stored in the same type of Dewar flask. Obviously care should be taken out to confuse the two, since serious explosion may result from the contact of liquid oxygen with organic matter.
9. Safety glasses of shields should be worm for all vacuum line work.
10. All Dewar flasks should be wrapped with tape.
11. All gas cylinders should be securely chained in place to avoid any possibility of upsetting.
12. Lead bricks and other heavy objects should not be stored in elevated positions.
13. Any operations in which high pressures might develop should be carried out behind explosion shields.
14. All moving parts of heavy equipment, such as vacuum pump motors, should be braced and protected there can be no danger of catching a person's clothing.

A. Working with Radioisotopes

It should be understood at the outside that there will be no hazard in working with radioisotopes, if the proper precautions and rules are observed. There is no need to be afraid of radiation, but there is a real need to appreciate its potential dangers, to learn safe radioisotope handling practices, to develop careful habits and to work methodically.

The personal rules are quite few

1. Never eat, drink or smoke when working with radioactive materials.
2. Always wear a laboratory coat. Wearing shoe covers is usually not necessary with the amounts of isotopes handled in the typical procedures in biological and agricultural application.
3. Always wear rubber gloves when working with radioactive materials. Wash and monitor them before taking them off carefully, without touching the outsides. Disposable polythene gloves are often used.
4. Wear a film badge or a pocket dosimeter.
5. Always use a pipeetting device (rubber suction bulb or plunger pipette when transferring aliquots of radioactive solutions. Never pipette by mouth.
6. Always work on a plastic tray lined with absorbent paper, when transferring radioisotope solutions or other active material liable to spill.
7. Use an adequate fume hood if there is likely to be vapour or fumes, e.g. from evaporation or ashing, which might be radioactive.
8. Return any stock radioisotope solution or sealed source to its place of storage immediately after use.
9. See that all containers, volumetric flasks etc. containing radioisotopes are adequately labeled, with details of the isotopes and its initial activity at the date of receipt or dilution.
10. It is normally permissible to handle planchets for "counting" without gloves, using forceps. Apply barrier cream such as "Savlon" to the hands before doing so.
11. After removal of rubber gloves always wash your hands with soap and after, and especially carefully before leaving the laboratory for meals at the end of the day.
12. Always work with a beta-gamma monitor beside you, and always monitor your hands, shoes, clothing (coat cutts etc.) at the end of the day, or after you have been handling relatively high activity material, as when diluting stock solutions.

13. Never take your rubber gloves into the counting room, and it is usually desirable not be wear the same laboratory coat in the counting room that is worm in the preparation laboratories. Failure to take care in this respect will soon raise the general background radiation level and hinder counting of low activity samples.

These personal routine procedures for safe working with radioisotopes have been given first so that the reader may better appreciate the discussions which follow on laboratory design, shielding, avoiding contamination, health and safely requirements, personal monitoring and waste disposal.

B. Laboratory Equipment

A listing of specialized equipment for a small to moderate-sized radioisotope laboratory follows:

Item	Suggested quantity	Remarks
1	2	3
Radio chemical fume hood	1 – 2	
Chemical fume hood	1 – 2	
Gloved box	1	Two or more may be needed if used instead of hoods.
Isotope storage pots	4	Depends on amount of gamma activity to be stored.
Laboratory cart	1 – 2	Handy for transport of shielded radioactivity.
Shielding bricks	20 – 80	Few needed if work is confined to beta emitters.
3- ft. tongs	3	
Various small tongs	10	
Pantograph or remote control pipetter	1	
Scalers	2 – 4	May need more if repair facilities are not handy.
Automatic sample changer	1	Only if money is to be traded for technician' time in counting large numbers of samples.
Extra Geiger-Muller counters	6	A variety should be maintained to anticipate needs.
Scintillation counter	1	A "Must" for gamma counting.
Internal counter	1	May need more than one if most of work is to be done with soft-beta emitters.

Contd...

Vibrating-reed electrometer and ionization chamber	1	Only if high-sensitively C^{14} assay are required.
Lead counter shield	1 - 2	Can use lead bricks of necessary to economize.
Cutie Pie (ion-chamber survey instrument)	1	
Portable Geiger-Muller survey instrument	1	Two of these preferable if finances permit.
Line-operated Geiger Muller survey instrument	1	
Pocket dosimeter	6	
Charging box for above	1	
Absorber sets	1	Cover range from 0 to 1500 mg/cm^2
Syringe shields	2	Lucite for beta emitters, lead for gamma emitters.

Detectors and associated equipment

Equipment	Description
Ionization chambers	Chambers with separate and sensitive electrometers : with incorporated Laurisen electroscopes : portable chamber and galvanometer arrangement.
Proportional and Geiger-Muller counter tube	Both may be constructed as sealed tubes as flow counters. Cylindrical tubes : thick walled for gamma rays : thin-walled for gamma and beta rays. End window counters with mica windows. Windowless flow counters for soft beta emitters.
Voltage supply	High voltage supplied by a regulated electronic circuit. Batteries and transistors for portable equipment. Current drain on ionization chambers and counters usually negligible : that for photomultiplier tube depends on resister stack supplying the dynodes.
Pulse amplifiers	For counting and for pulse height analysis. Most have : fast rise time and short resolving time : provisions for non-overloading : gain stability. Linearity of response needed/for pulse-height work.
Preamplifiers and cathode followers	All detectors are high-impendence devices and work bet into a high impedance input. Also where signal must be passed through a longer high capacitance (low impedance) cable.

Contd...

Scaling circuits (scalers)	Pulses from a detector, either direct of via an amplifier may be fed into a scaler which in turn drive a register at reduced rate. Requirements are : short resolving time : proper input sensitivity : proper output pulse for driving the register : dependability.
Timers	Various degrees of automation : either pre-set count with time indicated or pre-set time with count indicated.
Registers	Commonly digital, rotary, tubular devices which can be driven by an electrical pulse from the scaling unit. Also used are dial-type registers with faster response.
Rate meters	Circuits indicate history of the counting rate, with a length of time of integration time constant which can be set for different values. Rate meter-recorders for chromatogram scans and pulse-height analyzers.
Pulse height analyzers	In single channel types, signals received from a linear pulse amplifier. Analyzer produces a uniform output pulse for each pulse form the amplifier which lies in a certain voltage range setting of width equal the window and lower limit equal the base line. Output pulse drivers a scalar of rate meter-recorder. Pulse-height spectrums obtain by keeping a constant window and scanning the base line at a constant rate. Multiple-channel analyzers keep nearly all pulses and sort them into channels : the more channels, them into channels : the more channels, the more accurate the presentation of the spectrum.
Coincidence and anti coincidence circuit	Circuits have tow inputs and an output. Coincidence circuit gives an output pulse for coincident input pulses arriving within its resolving time. The anti coincidence circuit gives an output pulse for input pulses except when a coincidence occurs. Used in standardization, in low level, low background counting in measuring gamma-gamma coincidences, and liquid scintillation counting.
Other equipment	Shielding of lead, iron or other material to reduce background : gas supplies for flow counters : samples changing equipment : print-out equipment : scanning and recording equipment for in vivo mapping of activity in the body : and vacuum equipment for handling gas samples.

C. Instrumentation for Radiation Protection

Perhaps the most important point is to ensure that the instrument is operating properly before use. As a practical procedure, each instrument when first received should be checked at a standard distance from some source, and a record made of the reading. It will then take just a few seconds each day to check it against the same source so that any loss of sensitivity will be noted. After the use of two or three instruments, the order of magnitude of the readings can be anticipated.

Cutie pie

This instrument is battery-operated and has an ionization chamber and usually three scale ranges of 25, 250 and 2500 mr/hr. It usually has to 2 to 3 mg/cm^2 window, so that it is suitable for monitoring the soft-beta emitters such ac C^{14}, S^{35} and Ca^{45}. It is primarily of value at the higher levels of radiation, as indicated by the scale range, and is not suitable for low-level scanning for surface contamination.

Portable Beta-Gamma Count-rate Meter

This instrument is a batter operated Geiger-Muller counter and is useful when the radiation field is less than 20 mr/hr. Thus, in contrast to the Cutie Pie, it is ideal for scanning surfaces and is not suitable at the higher levels of radiation. At high levels it will "Jam" and show a zero reading, which must not be interpreted as indicative or zero radiation under such conditions. These instruments can be used with glass counter tubes of about 30-mg/cm^2 window thickness or thin-mica-window counters up to about 4 mg/cm^2. The latter must be used for the soft-beta emitters.

Line-operated Beta-Gamma Count-rate-Meter

This instrument performs the same function as the above and is similar except that it is not dependent upon batteries and it therefore not so portable. It is very convenient for scanning desk tops or material that can be brought to it. These instruments usually read in count per minute. In addition to survey work, they can often be used for special types of quantitative studies.

Pocket Dosimeter

This instrument is about the size of a fountain pen and is designed to be worm upon the person. It is self-reading, has a scale of 200 mr and gives the total dose received from the time of charging to the time of reading. An auxiliary charging unit is required which suffices for any number of dosimeters. It is sensitive only to gamma and high energy beta rays. The advantage of this instrument is that it can be read by the wearer at any time to give the dose received. It can be conveniently placed within the gloved palm of the hand for an estimation of the hand dose during a given operation.

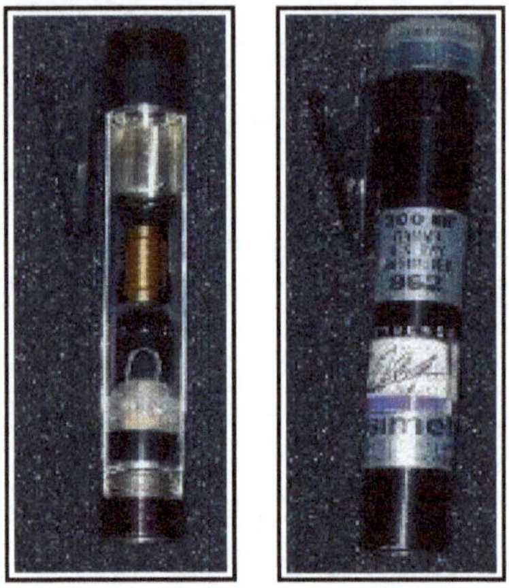

Fig. 1: Typical health-physics instruments

Ionization pocket meters of a condenser type are similar to the above but must to read an instrument. These may be more economical when large numbers of personnel are involved, but the dosimeter is more convenients. All these personnel meters have tendency for spurious reading due to spontaneous discharge, and it is advisable to wear two of them if possible.

Hoods

The purpose of the hood is to prevent contamination such as volatile compounds spray, or dust-from escaping into the laboratory. At low levels of the nonhazardous radioisotopes, work may be done on the open bench it is known that no air-borne contamination is being produced. Under slightly more hazardous conditions the ordinary chemical fume hood may suffice, although tests have shown that it is usually inefficient in preventing the spread of fumes into the laboratory. For general purposes, and in order to anticipate future work, it is desirable, if economically possible, to install a radiochemical fume hood, which is now available commercially. For work in the gas phase, particularly with carbon 14, a specially designed hood is desirable which permits ready access to a vacuum rack.

Gloved box

Experience has shown that the best of fume hoods cannot guarantee absolute freedom from laboratory contamination. In dealing with particularly hazardous materials, the gloved box should be used. A small blower maintains the box at a slightly reduced pressure, and the exhausted air may be filtered before being passed into the main hood stoack. For some types of work a simple box without a blower may be satisfactory. Depending upon the type and amount of work to be done, the gloved-box system may be used as an economical replacement for the fume hood.

Film badges

Film-badge service is commercially available. The developed film serves as a permanent record of the individual's exposure and may thus be of legal value. The film packet usually contains a sensitive and an insensitive film which cover the gamma dosage in the range of 30 to 20,000 mr. The primary disadvantage is the time lag between the exposure and the evaluation of the film. The film badge is also insensitive to soft-beta radiation. The value of the film badge is questionable in the small laboratory doing tracer work when adequate monitoring is available by other methods and detailed records are kept of instrument readings and amounts of activity.

Typical Floor Plans

It is convenient to discuss the laboratory layout in terms of areas set aside for specific purposes, which may or may not require separate rooms depending upon the magnitude of the experimental work.

The hot laboratory is used primary for unloading the radioisotopes and preparing the dilutions necessary for radiochemistry or for administration to the biological system. It should be located near the isotope-storage room, which may be used for storing both isotope shipments and radioactive wastes. Likewise the hot laboratory should be located near the injection area in which the radioisotopes are administered to animals. There is often the likelihood of spillage in this area. The hot animal or plant room is used to house biological specimens that have received radioactivity. The medium-level laboratory is used for dissection of active biological material, ashing procedures, and radiochemical operations at intermediate levels. At this stage care must be taken to avoid the cross contamination of samples, which will lead to counting errors. As the name implies, the low-level laboratory will be used for operations in which the materials have radioactivity at about the counting level. The counting room is used for the quantitative measurements of radioactivity in samples, and in this area stray radiation or radioactive contamination cannot be tolerated. Other facilities will be required for nonradioactive operation such as utility and storage rooms, a darkroom, animal or plant rooms, and offices: these present no special problems and will not be considered further.

Figure 2 illustrates a non-room laboratory of about 400 fit^2 that would be suitable for handling several millicuries of a soft-beta emitter or micro curie quanties of more energic emitters. The counting is done

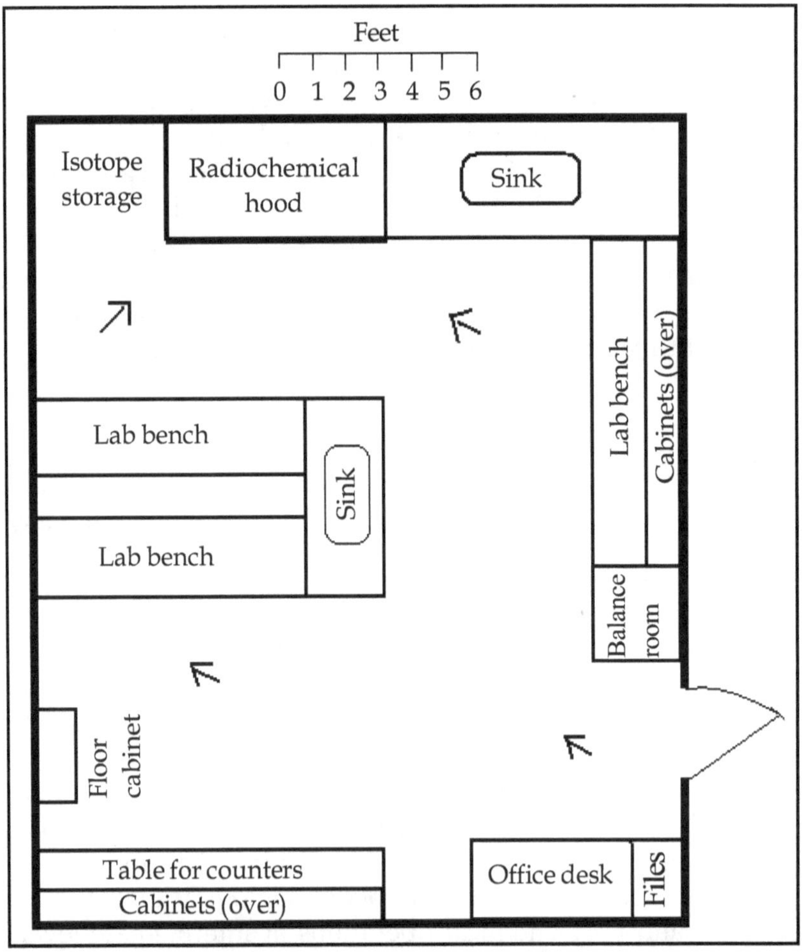

Fig. 2: Typical one room radiochemical laboratory of about 100 ft^2 (arrows indicate air-flow pattern)

as far as possible from the chemical operations, and the use of shielded counter tube is very helpful under these conditions. A two-room laboratory of about 700 ft^2 is shown in Fig.3.

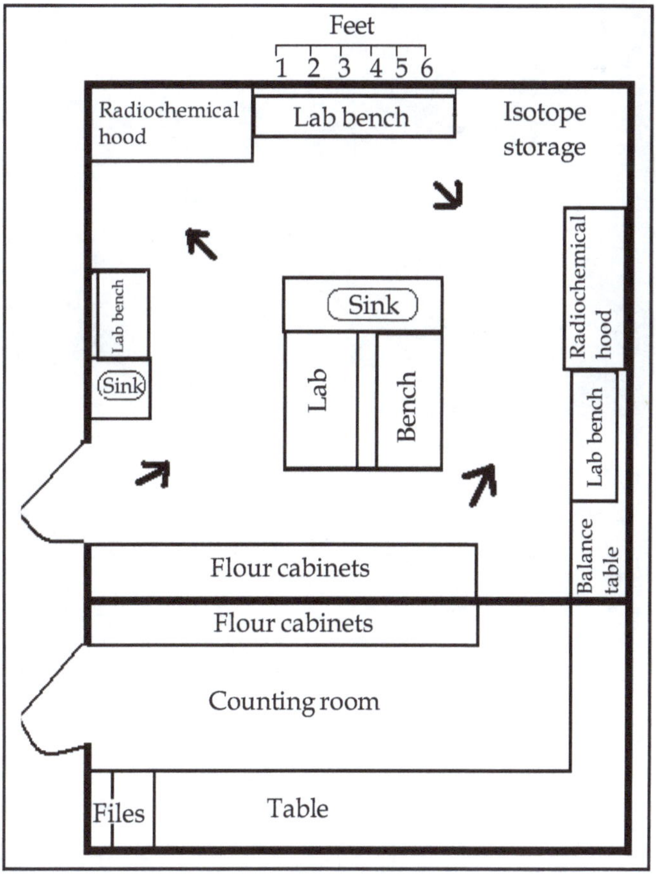

Fig. 3: Typical two-room radiochemical laboratory of about 700 ft^2

Glossary and Abbreviations

A_0 : Original activity.

Aao : Acid aluminum oxide, an inorganic ion-exchanger.

Absolute activation analysis : See activation analysis, absolute.

Absolute activity : See Activity.

Absolute counting : A measurement performed under such well-defined conditions that the absolute activity of a sample can be derived directly from the observed count rate.

Absolute full energy peak : See full energy peak efficiency, absolute.

Absolute photopeak efficiency : See photopeak efficiency, absolute.

Absorbed dose : See dose, absorbed.

Absorber : A substance used to absorb energy from any type of radiation. OR Any material that stops ionizing radiation. Lead, concrete, and steel attenuate gamma rays. A thin sheet of paper or metal will stop or absorb alpha particles and most beta particles.

Absorption (energy) : A phenomenon in which radiation transfers to matter which it traverses some or all of its energy.

Absorption coefficient (energy) : See absorption coefficient, linear.

Absorption cross section : For incident neutrons, the sum of all the microscopic or partial cross sections in which the incident neutron does not reappear as a product.

Absorption curve : A graph showing the relationship between the fraction of radiation absorbed and its energy, range, or wavelength.

Absorption edge : The energy or wavelength corresponding to a discontinuity in the absorption coefficient of a substance towards radiation .

Abundance ratio, isotopic : See abundance, isotopic.

Abundance, isotopic : The relative number of atoms of a particular isotope in a mixture of the isotopes of an element, expressed as a fraction of all the atoms of the element.

Abundance, natural isotopic : Of a specified isotope of an element, the isotopic abundance in the element as found in nature.

Accelerated test method : A procedure, preferably one for which a standard protocol has been developed, that is used in a laboratory to produce in a reasonably short time (days or months) effects that would otherwise be observed only over much longer times (decades to millennia). Generally this requires adjusting parameters such as temperature (and pressure), radioactive dose rate, etc., that affect the kinetics of a chemical reaction.

Accelerator : Device used to increase the energy of particles, which then collide with other particles. Major types are linear accelerators and circular accelerators. The name refers to the path taken by the accelerated particle. *OR* A device for generating a beam of high-velocity charged particles.

Accelerator mass spectrometry : A highly sensitive technique of mass spectrometry in which the sample, after ionization, is separated not only by mass but also by atomic number. To accomplish this, ions in the beam are accelerated to several MeV per mass number.

Acceptable limit : Limit acceptable to the regulatory body. See authorized limit.

Acceptor : In a solid state detector, an atom of a substance added to a semiconductor crystal to increase the number of holes in the conduction band.

Accountability : Quantitative accounting for nuclear material, inventories, and transfers, through a system of measurements, records and reports.

Acid digestion : In treating samples, the use of acid to chemically decompose a material into its simpler constituents (usually soluble) thereby releasing the analytes for subsequent analysis.

Actinide : An element which is any of the second (5f) series of f-block elements or inner transition elements commencing with actinium at atomic number 89 and ending with lawrencium at atomic number 103.

Actinium series : The naturally occurring decay chain commencing with ^{235}U (formerly known as actino-uranium) and with ^{207}Pb as the stable end product. The mass number of each radioactive member of the actinium series is representable as 4n+3 *where n* is an integer.

Actinoids : See actinides.

Actinometer : Any instrument, based on observable chemical reaction, used to measure the intensity of radiant energy.

Activated sludge : Term sometimes used with reference to radioactive effluent treatment processes employing biological slimes and sludges to concentrate the radioactive materials.

Activation : The process of inducing radioactivity by irradiation. In general, a specification is

		added of the type of incident radiation (e.g., nuclear, neutron, photon) or its energy (e. g. thermal, fast).
Activation analysis	:	A method of elemental or isotopic analysis based on the measurement of characteristic radiation) from nuclides formed directly or indirectly by activation). In general, a specification is added of the type of the incident radiation (e. g. neutron, proton) and its energy (e. g. , thermal, fast).
Activation analysis, post	:	Synonymous with activation analysis, post , which is preferred.
Activation analysis, absolute	:	A method of activation analysis in which the elemental concentrations in the material are calculated from known nuclear constants, irradiation and measurement parameters, rather than by comparing with known standards.
Activation analysis, instrumental	:	A method of activation analysis in which element specificity is obtained by using appropriate irradiation conditions, radiation measurement techniques and mathematical techniques for the interpretation of the measurement results.
Activation analysis, neutron	:	See activation analysis.
Activation analysis, non-destructive	:	An activation analysis procedure in which after the irradiation , no chemical and physical operations are applied which cause a change of any properties of the sample.
Activation analysis, radiochemical	:	A method of activation analysis in which, after the irradiation , chemical or physical separation is applied.
Activation cross section	:	See cross section, microscopic.
Activation detector	:	A radiation detector in which the induced radioactivity produced by exposure in a radiation field is used to determine particle flux density or particle fluence.

Activation foil : An activation detector in the form of a foil.

Activation product : An element made radioactive by bombardment with charged particles, photons, or neutrons.

Active (neutron) assay or interrogation : A non-destructive measurement technique in which the sample is exposed to (neutron) irradiation which causes a reaction (usually nuclear) that can be measured.

Active deposit : The non-gaseous, radioactive descendants of radon from the natural radioactive series (decay chain).

Activity : The rate of radioactive decay. ***OR*** The number of nuclear decays occurring in a given quantity of material in a time interval, divided by that time interval. Synonymous with disintegration rate.

Activity concentration : For a specified radionuclide , the activity of a material divided by its volume.

Activity curve : See decay curve .

Activity, molar : For a specified isotope in a compound, the activity of one mole of the compound.

Activity, specific : For a specified radionuclide , the activity of a material divided by the mass or volume of that material.

Adc : Analog-to-digital converter.

Adsorbent : A material that has the property of adsorbing radionuclides (see adsorption).

Adsorption : The enrichment of one or more components in an interfacial layer.

Aes : Auger electron spectroscopy.

Age, neutron : See Fermi age.

Age, radioactive : See radioactive age.

Agonist : A chemical which can not only combine with a receptor , like an antagonist , but when it does so stimulates it, resulting in an observable effect.

Air equivalent : A measure of the effectiveness of an absorber of radiation equal to the thickness of a layer of air at STP that absorbs the same fraction of radiation or results in the same energy degradation as the air.

Airborn debris or waste : Radioactively contaminated matter, generally small solid particles or liquid droplets that have become suspended by the air.

Alara : As low as reasonably achievable, in reference to radiation hazards.

Albedo : The reflection factor a surface, such as paraffin, has for neutrons.

Ali : Annual limit of intake.

Alpha : See alpha particle .

Alpha particle (alpha radiation, alpha ray) : A ^{4}He nucleus. It is made up of two neutrons and two protons. It is the least penetrating of the three common forms of radiation, being stopped by a thin sheet of paper. It is not dangerous to living things unless the alpha-emitting substance is inhaled or ingested or comes into contact with the lens of the eye.

Alpha-decay : Radioactive decay in which an alpha particle is emitted.

Alpha-emitter : A nucleus that undergoes radioactive decay by emitting an alpha particle .

Alpha-particle : Alpha-particle. A ^{4}He nucleus emitted during a nuclear transformation. Also called an α-ray or just an alpha.

Amalgam exchange : A rapid and efficient separation procedure for a number of cations from aqueous solution into a mercury amalgam of their elements.

Amplifier, biased linear pulse : A pulse amplifier which, within the limits of its normal operating characteristics, has a constant gain for that portion of an input pulse that exceeds the threshold value and that produces no output for pulses whose amplitude is below the threshold.

Amplifier, linear pulse : A pulse amplifier which, within the limits of its normal operating characteristics, delivers an output pulse of amplitude proportional to that of the input pulse.

Ams : Accelerator mass spectrometry .

Amu : Atomic mass unit

Analog to digital converter (ADC) : A pulse amplitude analyzer which produces an digital number proportional to the amplitude of the pulse presented at its input. The three types are the linear ramp converter (Wilkinson type), the successive approximation ADC, and the flash ADC.

Analyte : The substance in a specimen to be analyzed in an assay.

Analytical radiochemistry : See radioanalytical chemistry .

Angular correlation : In reference to radiation in coincidence, the dependence of the direction of one radiation with respect to its coincident partner.

Anion exchange : The process of exchanging anions between solution and an anion exchanger .

Anion exchanger : An ion exchanger with anions as counter-ions.

Annealing : In radiation of materials, the process of returning displaced atoms to their original position by heating or irradiation with electrons or gamma rays. The original displacement can have been caused by a variety of processes including recoils from nuclear transformations.

Annihilation : An interaction between a particle and its antiparticle in which the pair is converted into annihilation photons.

OR Annihilation of particles is the disappearance of the mass energy of a particle and its corresponding antiparticle, and its appearance as another sort of energy (possibly including a spray of particles of total quantum number zero for each of the additive quantum numbers).

Annihilation photons : Photons produced as the result of annihilation of a particle with its antiparticle . The summed energy of the annihilation photons must equal the mass-energy equivalent of the summed masses of the annihilating particles.

Annihilation radiation : Radiation resulting from annihilation.

Annual limit of intake : Annual limit of intake, in reference to radiation dose. The activity of a radionuclide that, taken into the body during a year, would provide a committed effective dose equivalent to a person equal to the annual occupational effective dose equivalent limit or, in some cases, the organ dose equivalent limit.

Antagonist : A substance that binds to a receptor but does not activate it, in the process of blocking the binding of the natural agonist and so is preventing its action.

Antibody : A serum or cell-bound protein, part of which provides the specificity for binding an antigen .

Anticoincidence : The occurrence of an event that is unaccompanied by one or more other specified events at the same time; equivalent to a Boolean "not".

Anti-compton gamma-ray spectrometer : See gamma-ray spectrometer, anti-Compton.

Antigen : Any substance that will induce antibody formation after parenteral application to a vertebrate animal.

Antineutrino : The antiparticle of the neutrino, emitted with the negative beta particle in beta decay. *OR* Antiparticle to the neutrino. See antiparticles.

Antiparticle : Of a given particle , a particle with the same rest mass and opposite charge; the two can undergo annihilation. *OR* Particle having the same mass, spin, isospin as a particle, but having all additive quantum numbers opposite to those of its respective particle. Antiparticles have the opposite charge of its corresponding particle. Antibaryons are antiparticles to baryons, antileptons are antiparticles to leptons, antiquarks are antiparticles to quarks. The antiparticle for a particular particle, for example a neutrino, is denoted an antineutrino.

Appearance potential : The minimum potential which the electron beam in an ion source must traverse in order to acquire enough energy to produce ions of a specified nuclide or molecular fragment.

Archaeometry : The application of the measuring techniques of the hard sciences to the remains left by ancient peoples.

Artificial element : An element which has been produced through nuclear transformation and which does not otherwise exist in nature.

Artificial radioactivity : See induced radioactivity.

Ashing : Dry or wet mineralization as a method of preconcentration of trace substances.

Assay : 1. The quantitative determination of a specified analyte . 2. The measurement of the quantity of a chemical constituent, or of the activity of an enzyme, hormone or other biological compound, or biological potency of a drug in a biological material.

Assay kit : A set of components (reagents and other necessary materials) and procedural instructions packaged together and designed for the estimation in vitro of a specified analyte, when used according to the instruction.

Assay meter : An instrument used for the assay of radioactive material. It can comprise a radiation detector sensitive to the emissions from the radioactive material and a counting rate-meter or a scaler .

Asymmetry : Denoting absence of symmetry. Sometimes a quantitative measure of a peak so characterized.

Asymptotic freedom : Quark-quark interactions weaken as the energy gets higher, or, equivalently, as the quarks approach one another.

Atm : Atmospheres of pressure

Atom : The smallest unit quantity of an element that is capable of existence whether alone or in chemical combination with other atoms of the same element or other elements. *OR* A particle of matter indivisible by chemical means. It is the fundamental building block of molecules. It consists of a positively charged nucleus and orbiting electrons. The number of electrons is the same as the number of protons in the nucleus.

Atomic mass (sometimes mistakenly called atomic weight) : Rest mass of an atom in its nuclear and atomic ground state. *OR* The mass of a neutral atom. Its value in atomic mass units (u) is approximately equal to the sum of the number of protons and neutrons in the nucleus of the atom.

Atomic mass number : A, the total number of nucleons (protons and neutrons) found in a nucleus.

Atomic mass unit : A special unit of mass, equal to 1/12 of the rest mass of an atom of nuclide ^{12}C in its nuclear and atomic ground state .

Atomic mass unit (amu or u) : Unit of mass defined by the convention that the atom ^{12}C has a mass of exactly 12 μ; the mass of 1 μ = 1.67×10^{-27} kg.

Atomic mass, relative : The ratio of the mass per atom of an element, averaged over a specified nuclidic composition, to 1/12 of the mass of an atom of nuclide ^{12}C. The term atomic weight is also used for this quantity.

Atomic number : Number of protons contained in a nucleus. Synonymous with proton number.

Atomic number : Z, the total number of protons found in a nucleus.

Atomic pile : An obsolete synonym for a nuclear reactor.

Atomic weight : Atomic mass.

Attenuation : The reduction of a radiation quantity upon its passage through matter. Attenuation results from interactions of the radiation with the matter it traverses.

Attenuation factor : A measure of the opacity of a layer of material for radiation traversing it; the ratio of the incident intensity to the transmitted intensity.

Auger (electron) yield : The fraction of the atoms having a vacancy in an inner orbital which relaxes by emission of an Auger electron.

Auger effect : The emission of an electron from an atom ; accompanies the filling of a vacancy in an inner electron shell.

Auger electron : Electron originating in the Auger effect.

Auger electron spectroscopy : Any technique in which a specimen is bombarded with keV-energy electrons or X-rays, and the energy distribution of the electrons (Auger electrons) produced through radiationless de-excitation of the atoms in the sample is recorded.

Authorized limit : Limit set for a given radionuclide or source or for a given environment by a national or international environmental authority.

Auto-ionization : The radiationless transition from an atom to an equal energy level in the continuum of an ion.

Autoradiogram : See autoradiograph.

Autoradiograph : A radiograph of an object containing radioactive substance, produced by itself when it is placed upon photographic plate or film.

Autoradiography : A collection of techniques whereby radioactivity is visually localized within solid specimens by means of the close apposition of a detector layer.

Autoradiography, whole body : Autoradiography of cross-sectional slices.

Autoradiolysis : Radiolysis of a radioactive material resulting directly or indirectly from its radioactive decay.

Average life : The average life of an atom or nuclear system in a specified state. For an exponentially decaying system, it is the average time for the number of atoms or nuclei in a specified state to decrease by a factor. Synonymous with mean life.

Average logarithmic energy decrement : In moderation of neutrons, it is the average of the natural logarithm of the ratio of the energy of a neutron before a collision to that after a collision.

Back extracting : See stripping.

Background : The term employed to designate the value indicated by a radiation measuring device in the absence of the source whose radiation is to be measured, when the device is placed under its normal conditions of operation.

Background radiation : Radiation from any source other than the one to be detected or measured. *OR* The radiation found in the natural environment originating primarily from the naturally radioactive elements of Earth and from cosmic rays. The term may also mean radiation extraneous to an experiment.

Backscatter : Scattering of radiation in a generally backward direction. In the assay of radioactivity , it applies to the scattering of radiation into the radiation detector from any material except the sample and the detector.

Backwashing : In an ion-exchange column, an upward flow of eluant through the resin bed serving to clean and reconstitute the exchanger bed.

Badge : A dosimeter that is clipped onto clothing or worn as a "ring" on the finger and which is later processed to quantify exposure to radiation.

Ballistic deficit : In proportional counters , when pulses are shaped using time constants adjusted so that the slow component of the drift of ions do not contribute to the pulse amplitude, the shaped pulse has a lower amplitude than that corresponding to an infinite time constant by an amount known as the "ballistic deficit".

Barn : A unit of area used in expressing nuclear cross sections : 1 barn = $10^{-28}m^2$.(Not a SI unit. Symbol is "b".)

Baryon : A massive composite hadron (made of three quarks) such as the proton or the neutron.

Baryon number : Quantum number characteristic of baryons. Each baryon has a value of +1, while each anti-baryon has a value of -1.

Batch process : A process, particularly in reference to separations, that is not continuous and which is conducted with discrete quantities of materials or a limited number of reagents.

Bateman equations : The set of coupled differential equations that expresses the amounts of reactants and products as a function of time as these species are undergoing both production and loss by nuclear reaction and radioactive decay.

Beam : Collimated radiation for the purpose of irradiation .

Beam current : In a charged particle accelerator, the electric current determined by the number, velocity, and charge of the particles.

Beam dump : See beam stop.

Beam hole : In a reactor, a hole through the shielding which allows a beam of radiation, especially fast neutrons, to escape.

Beam monitoring : Measuring the intensity or energy flux density of a beam .

Beam stop : Material at the end of a beam placed there to completely stop and absorb the remaining beam particles .

Beam trap : See beam stop

Becquerel : SI unit of activity or nuclear transition rate equal to one per second (symbol 1 Bq). Caveat illustration: The activity of 1 nanomole of 40K (half life = 1.25×10^9 years) would be approximately 10.5 Bq but the gamma activity of that same nanomole would be only 1.13 Bq because only 10.7% of 40K's decays involve a gamma branch. To avoid ambiguity, the latter should be referred to as 1.13 gamma Bq.

Becquerel (bq) : Unit of activity in the International System– one disintegration per second; 1 Bq = 27 pCi.

Beta decay : Nuclear decay in which a beta particle is emitted or in which orbital electron capture occurs.

Beta emitter : An atom or nucleus which undergoes radioactive decay by emitting a beta particle (β) and an antineutrino or neutrino. The former accompanies a $\hat{a}^-$ and the latter a β^+.

Beta particle : A negative or positive electron which has been emitted by a nuclear particle in a nuclear transformation.

Beta particle (beta radiation, beta ray) : An electron of either positive charge (e^+) or negative charge (e^-) emitted by an atomic nucleus or neutron in the process of a transformation. Beta particles are more penetrating than alpha particles but less than gamma rays or x-rays. Electron capture is a form of beta decay.

Beta stability line : On a graph of neutron number or of mass number vs atomic number and displaying one or more properties of each nucleus, the locus of stable nuclei. (Synonym: Segre curve)

Bgo detector : A solid scintillation detector material based on bismuth germinates.

Biased linear pulse amplifier : See amplifier, biased linear pulse.

Bifunctional chelate : Complexing agent with two sites for complexation.

Big bang : Beginning of the universe; a transition from conditions of unimaginable density and temperature to conditions of lower density and temperature.

Bi-isotopic monitor : The use of ^{95}Zr and ^{97}Zr for determination of the thermal to epithermal neutron flux ratio by activation analysis .

Binding : In competitive radio assay, the reactive forces between ligand and binding agent describable by the mass action law and measurable by the fraction of reagent tracer bound in the ligand-binding agent complex.

Binding agent : In competitive radio assay , the test reagent chosen, most commonly antibody, to react specifically with the substance under test via mass action, reversible reaction.

Binding capacity : The amount of specific binding sites available per quantity of binding reagent. The extrapolated point of the X-axis of a Scatchard plot performed under defined conditions.

Binding energy : The energy necessary to break a nucleus into specified constituents; the total binding energy is the energy requirement to disassemble the nucleus entirely into protons and neutrons. The binding energy of a neutron in a nucleus is synonymous with its separation energy from that nucleus. It is the energy equivalent of the difference between the sum of the masses of the constituent neutrons and protons and the measured mass of that nucleus.

Bioconjugate : An agent (usually a chelate used to conjugate radionuclide to an antibody.

Biological half-life : For a substance the time required for the amount of that substance in a biological system to be reduced to one half of its value by biological processes, when the rate of removal is approximately exponential.

Black hole : An object so dense that light cannot escape from it.

Blackbody : An object that is a perfect emitter and absorber of radiation.

Blackbody radiation : Radiation emitted by a blackbody (the intensity depends on temperature).

Bnct : Boron neutron capture therapy.

Boiling water reactor : A nuclear power reactor with enriched ^{235}U fuel and boiling light water as moderator and coolant.

Bolus : A contiguous mass, such as a medically administered dose of labeled compound.

Bombardment : Irradiation in a beam of particles or photons.

Boron carbide filter : In epithermal neutron activation analysis, a thickness of boron carbide (B_4C) that is used instead of the more usual cadmium filter which is used to achieve the cadmium cutoff . In this way, one may avoid the induced activity that grows into the cadmium with use.

Boron chamber : An ionization chamber that is lined or filled with boron or boron compounds. M.

Boron nitride filter : In epithermal neutron activation analysis, a thickness of boron nitride (BN) can replace the more usual cadmium filter used to achieve the cadmium cutoff. In this way, one may avoid the induced activity that grows into the cadmium with use.

Boson : A particle having spin that is an integer multiple of (h-bar).

Bowen's standard kale : A reference material for the analysis of plant materials with known concentration values and ranges for a group of macro elements (Na, K, Ca, Mo, O, P, S, N, and Cl) as well as 30 trace elements.

Bq : The symbol for the becquerel unit of activity of a radioactive substance. *OR* Becquerel.

Bragg curve : A graph of specific ionization due to a charged particle passing through a material as a function of the penetration depth of that particle relative to its range; that is, as a function of its residual range .

Bragg peak : A peak in a Bragg curve showing that the amount of ionization per millimeter of tissue or other stopping medium traversed by charged particles increases sharply as the particles slow down near the ends of their tracks.

Bragg rule : The mass stopping power of a compound or mixture for charged particles is approximately given by the weighted sum of the stopping powers of all the components atoms.

Branching decay : Nuclear decay which can proceed in two or more different ways.

Branching fraction : In branching decay the fraction of nuclei which decay in a specified way. Synonymous with branching probability.

Branching probability : See branching fraction.

Branching ratio : The ratio of the branching fraction for two or more specified modes of decay.

Breit-wigner formula : A formula describing the rapidly varying cross section for a nuclear reaction in the vicinity of one or more resonances .

Bremsstrahlung : The electromagnetic radiation associated with the acceleration or deceleration of charged particles in the presence of a Coulomb field.

Bremsstrahlung, inner : Bremsstrahlung which may accompany the emission or absorption of a charged particle by a nucleus.

Build-up factor : In the passage of radiation through a medium, the ratio of the total value of a specified radiation quantity at any point to the contribution to that value from radiation reaching the point without having undergone a collision.

Burial ground : A place for burying unwanted radioactive objects to prevent their escape or escape of their radiations.

Burn-up : Induced nuclear transformation of atoms during reactor operation.

Burn-up fraction : The fraction of an initial quantity of a given nuclide that has undergone burn-up .

Burn-up, specific : The total energy released through induced nuclear transformations divided by the mass of a nuclear fuel.

Burst : A sudden increase or appearance of particles or fragments emitted, of short duration, as in a nuclear bomb explosion.

Bwr : See boiling water reactor.

By-product : Any radioactive material (except special nuclear material) yielded in or made radioactive by exposure to the radiation incident to the process of producing or utilizing special nuclear material. The tailings or waste products produced by the

		extraction or concentration of uranium or thorium from any ore processed primarily for its source material content.
Ca	:	Carrier added.
Cadmium cutoff	:	In neutron irradiations, the energy value which, for a given experimental configuration, is determined by the condition that if a cadmium cover surrounding a detector were replaced by a fictitious cover black to neutrons below this value and transparent to neutrons with energy above this value, the observed detector response would be unchanged.
Cadmium ratio	:	The ratio of the response of an uncovered neutron detector to that of the same detector under identical conditions when it is covered with cadmium of a specified thickness.
Cage effect	:	See clathrate .
Camac module	:	A standardized modular instrument compatible with a bin or crate that houses and powers the instrument and that also has a built-in digital data bus to provide computer communications with the unit.
Candu	:	Canadian heavy-water moderated nuclear reactor using natural uranium as fuel.
Capture	:	A process in which an atomic or nuclear system acquires an additional particle. In general a specification is added of the type of the captured particle or its energy.
Capture cross-section	:	The cross section for capture .
Capture gamma radiation	:	The gamma radiation emitted in radioactive capture.
Capture, electron	:	A nuclear transformation in which the nucleus captures an orbital electron . Often the shell from which the electron is captured is indicated. (K, L, etc.)

Capture, radiative : Capture of a particle by a nucleus followed by immediate emission of gamma radiation.

Capture, resonant : Capture of a particle exactly matching the product's resonance energy.

Carrier : An inactive material deliberately added to a specified radioactive substance to ensure that the radioactivity will behave normally in all subsequent chemical and physical processes.

Carrier proteins : Macroscopic amounts of nonlabelled proteins present with trace amounts of radiolabelled proteins.

Carrier, hold-back : A non-radioactive carrier of a radioactive nuclide used to prevent that particular radioactive species from contaminating other radioactive species in a chemical operation.

Carrier, isotopic : A carrier which differs only in isotopic composition from the trace substance it has to carry.

Carrier, nonisotopic : A carrier which is not an isotope (that is, it is a different element) of the trace substance or substances it has to carry. A scavenger of radioactive elements.

Carrier-free : A preparation of a radioactive isotope which is 'free' from stable isotopes of the element in question. More precisely, a preparation of a radioactive isotope of high specific activity to which no isotopic carrier was intentionally added and which was not produced by irradiation of stable isotope of the same element.

Cascade, auger : In the Auger effect, when the emission of an Auger electron leads to the creation of a new vacancy that relaxes by a subsequent Auger effect.

Cascade, gamma : See gamma cascade.

Catcher foil : In an irradiation of a target, a foil used to stop and retain nuclear reaction products that recoil out of the target.

Cation exchange : The process of exchanging cations between solution and a cation exchanger .

Cation exchanger : An ion exchanger with cations as counter-ions.

Cba : See competitive binding assay.

Cbf : Cerebral blood flow

Cdd : Concentration dependent distribution.

Cell, hot : See hot cell.

Cema : An intermediate dosimetric quantity: the converted energy, per unit mass, expended by charged primary particles which in turn had been produced by uncharged particles.

Cems : Conversion electron Mssbauer spectroscopy.

Cerenkov detector : A charged particle detector based on the Cerenkov effect .

Cerenkov effect : Emission of radiation in the visible and ultraviolet spectrum arising when a charged particle crosses a medium with a velocity greater than that of light in the same medium..

Cerenkov radiation : Radiation resulting from the Cerenkov effect .

Cf : Carrier-free.

Cgy : CentiGray

Chain fission yield : The fraction of fissions giving rise to fission fragments of a particular mass number.

Chain reaction : In nuclear reactions, a process in which a reaction produces more than one reaction product (such as neutrons) that induces subsequent nuclear reactions.

Characteristic curve (of a GM counter) : The dependence of the response of a (Geiger-Muller) detector on the applied voltage. GM counters are operated in the "plateau" voltage region of the characteristic curve where the count rate is essentially unaffected by small changes in the operating voltage.

Characteristic x-radiation : X-radiation consisting of discrete wavelengths which are characteristic of the emitting element.

Charge exchange : In the interaction between two particles at least one of which is charged, the exchange of charge by those particles.

Charge trapping : In a semiconductor detector, once electrons and holes are formed by radiation, they can become immobilized for relatively long periods by impurities in the crystal lattice and this "trapping" can prevent the charge carrier from contributing to the measured pulse.

Charged-particle activation analysis : Nuclear activation analysis in which the irradiation is done using charged particles such as protons or alphas.

Chelate : A chemical which can undergo or has undergone chelation with an element.

Chelating resin : An ion exchange resin in which the counter-ions are retained by a chelate functional group on the resin.

Chelation : In molecular or complex ion structure, the formation or presence of bonds (or other attractive forces) from two or more separate binding sites within the same legend to a single central atom.

Chemical dosimeter : A dosimeter in which the dose is measured by observing the extent, under specified conditions, of a chemical reaction caused by the ionizing radiation to be measured.

Chemical isotope exchange : Exchange of isotopes of a given element between different molecules or between different chemical forms of this element in the course of a chemical reaction.

Chemical neutron activation analysis (cnaa) : A variation of activation analysis where a single element or group separations are employed prior to irradiation. Synonymous with activation analysis with preconcentration .

Chemical yield : The fraction of the amount of an element or chemical compound (remaining) following a specified chemical reaction or separation.

Chemiluminescence : Emission of light as a result of a chemical reaction in which there is no apparent change in temperature.

Chemisorption : Adsorption in which the forces involved are valence forces of the same kind as those operating in the formation of chemical compounds.

Chemistry, nuclear : The scientific discipline dealing with understanding and applying effects of a chemical and physicochemical nature caused by the nucleus of an atom, its transformations through decay and reactions, and nuclear radiation.

Chemistry, radiation : See radiation chemistry.

Chemonuclear reaction : A chemical reaction induced by nuclear radiation or fission fragments.

Cherenkov radiation : Light emitted by particles that move through a medium in which the speed of light is slower than the speed of the particles.

Chloramine-t : A common oxidizing agent used in radioiodination of proteins. N-chloro-4-methylbenzenesulfonamide sodium salt.

Chopper : A mechanical device, conceptually like a camera shutter, for controlling the precise timing of a beam .

Chromatogram : A graphical or other representation of a detector response, effluent concentration, or other quantity used as a measure of effluent concentration versus time or the volume of the effluent.

Chromatograph : 1.) To separate by chromatography. 2.) The assembly of apparatus for carrying out chromatographic separations.

Chromatography : A method used primarily for the separation of the compounds of a sample in which components are distributed between two phases, one of which is stationary while the other moves.

Ci : Symbol for the curie , a unit of activity equal to 3.7 X 1010 becquerels. OR curie

Cinaa : Cyclic instrumental neutron activation analysis. See cyclic activation analysis .

Cladding : An external layer of material (usually of Zircalloy, stainless steel, magnesium), directly surrounding nuclear fuel or other substance that seals and protects it from the environment and protects the environment from radioactive material produced during irradiation.

Clathrate : An inclusion compound; that is, a complex in which one compound (the host) forms a crystal lattice containing spaces in the shape of long tunnels or channels in which molecular entities of a second chemical species (the guest) are located. If the host spaces are closed on all sides so that the guest is "trapped", such compounds are known as clathrates or "cage compounds".

Cm : Centimeter(s)

Cmcp : Coincident measurement of complementary particles.

Cnaa : Chemical neutron activation analysis.

Co : Cobalt

Cockroft-walton generator : An electrostatic device for generating high voltages through staged rectifiers and transformers. The generators are used to accelerate charged particles.

Cocktail : (a) In nuclear medicine; ingested mixture containing radioactive or isotopic tracers for medical studies. (b) in liquid scintillation counting, a mixture of organic solvents and fluors into which the sample is dispensed for counting.

Cocrystallisation : See Coprecipitation.

Coincidence : The occurrence of two or more events at the same time.

Coincidence circuit : An electronic circuit which produces a usable output pulse only when each of its inputs receives a pulse within a specified time interval.

Coincidence correction : A correction to the count rate of a radiation due to losses or gains when two radiations arrive at and are processed by the detector within the coincidence resolving time .

Coincidence resolving time : The greatest time interval that can elapse between the occurrence of two or more consecutive signal pulses, in order that the measuring device processes them as a coincidence .

Coincidence, delayed : The occurrence of two or more events separated by a short but measurable time interval.

Coincidence, prompt : The occurrence of two or more events separated by a time interval which is less than a specified small value.

Coincidence, random : A coincidence of events occurring in different atoms or between measured atoms and other atoms, including background and unknown sources.

Coincidence, true : A coincidence of events triggered by a single cause. (Contrast with random coincidence above.)

Coincident measurement of charged particles : A technique used in charged particle activation analysis of light elements in thin targets in which an emitted charged particle is detected in coincidence with the recoiling product nucleus.

Cold test : A test of method, process, apparatus, or instrumentation with highly radioactive materials replaced by inactive materials or materials that may contain radioactive tracers.

Collateral series : A radioactive decay series, initiated by transmutation, that eventually joins one of the four natural actinide decay chains. Also called a collateral chain.

Collection efficiency : The chemical yield of a particular collection step in a separation, the step being either chemical or physical in nature.

Collectron : A neutron detector in which an electric current is produced without the application of an external power source though the emission of á particles by a short-lived radionuclide. This radionuclide is produced by neutron activation in the part of the detector called the "emitter".

Collimation : The limiting of a beam of radiation to the required dimensions and angular spread.

Collimator : An arrangement of absorbers used for collimation.

Collision density : In neutron transport theory, the number of collisions per cm3/s undergone by a neutron with energy E, equal to the neutron flux divided by the scattering mean free path.

Column (in chromatography) : The tube that contains the stationary phase and through which the mobile phase passes.

Column generator : See radioisotope generator.

Comparative lifetime : See ft value.

Comparator : A known amount of an element, capable of being activated, that is simultaneously irradiated with the test portion in the context of activation analysis . If one comparator is used (single comparator method) , it is essentially identical to a flux monitor (except that this term is not necessarily linked to activation analysis).

Competitive binding assay : Assay based on the competition between a labeled and an unlabeled ligand in the reaction with a receptor binding agent (e.g. antibody, receptor, transport protein).

Competitive protein-binding assay : A competitive binding assay in which the binding agent is a protein.

Complex : A molecular entity formed by a loose association involving two or more component molecular entities (ionic or uncharged) or the corresponding chemical species. The binding between the components is normally weaker than in a covalent bond.

Compound nucleus : A transient nucleus formed when two nuclear particles fuse completely; analogous to the activated complex in the transition state theory of chemical reactions.

Compton edge : In a gamma-ray spectrum, the maximum energy deposited by gamma-rays which are scattered by the Compton effect and consequently deposit less than the full energy peak . A continuum of measured energies is detected below the edge determined by the Compton effect.

Compton effect : The inelastic scattering of a photon by a free or weakly bound electron which afterwards occurs as a free electron. Part of the energy and momentum of the incident photon is transferred to the electron and the remaining part is carried away by the scattered photon. Synonymous with Compton scattering.

Compton electron : The energetic electron resulting from the Compton effect.

Compton scattering : See Compton effect.

Compton shift : The change in wavelength or energy of scattered radiation due to the Compton effect.

Compton suppression spectrometer : See gamma-ray spectrometer, anti-Compton.

Conservation law : A relation asserting that a specific quantity is conserved. For example, conservation of energy, conservation of momentum, conservation of electron number.

Containment : Any procedure which prevents the transport of radioactivity .

Contamination : The presence of an unwanted radioactivity in a material or the environment in a concentration exceeding the required limit or natural background.

Conversion coefficient, internal : For a transition between two specified energy levels of a nuclide , the ratio of the transition probabilities for internal conversion and gamma ray emission.

Conversion electron : Electron ejected from the atom in the process of internal conversion .

Conversion, internal : A transition between two energy states of a nucleus where the energy difference is given to an orbital electron which is thereby ejected from the atom .

Cooling, radioactive : Of a strongly radioactive material, the decrease of its activity by nuclear decay.

Coordination : The formation of a covalent bond, the two shared electrons of which have come from only one of the two parts of the molecular entity linked by it, as in the reaction between a Lewis acid and a Lewis base to form a Lewis adduct; alternatively, the bonding formed in this way.

Coprecipitation : The simultaneous precipitation of a normally soluble component with a macro-component from the same solution by the formation of mixed crystals, by adsorption , occlusion, or mechanical entrapment. (Applies generally to a radionuclide at the tracer level.)

Cosmic rays : High-energy particles irradiating the earth, including electrons, muons, protons, alphas, and heavy-ions, which originate extra terrestrially from the sun and from galactic and extra-galactic sources.

Coulomb barrier : The repulsive potential energy between two charged particles at a separation distance corresponding approximately to contact.

Count : 1. Information corresponding to a pulse processed for counting. 2. Number of pulses recorded during a measurement.

Counter tube : Radiation detector consisting of a gas-filled tube or valve whose gas amplification is much greater than one, and in which the individual ionizing events give rise to discrete electrical pulses. Often an expression is added indicating the geometry (e. g. end window), composition of the gas (e. g. helium) or the physical process essential for its operation (e. g. proton recoil, fission).

Counter tube, geiger-mller : A counter tube operated under such conditions that the magnitude of each pulse is independent of the amount of energy deposited in it.

Counter tube, proportional : A counter tube operated under such conditions that the magnitude of each pulse is proportional to the amount of energy deposited in it.

Counter, radiation : A radiation detector which measures a count rate but does not perform energy spectrum analysis .

Counting efficiency : The ratio between the number of particles or photons counted with a radiation counter and the number of similar particles or photons emitted by the radiation source .

Counting loss : A reduction of the counting rate resulting from phenomena such as the resolving time or the dead time.

Counting rate : The number of counts occurring in unit time.

Counting, absolute : A measurement under such well-defined conditions that the activity of a sample can be derived directly from the observed counting rate.

Cow : See radioisotope generator .

Cpaa : Charged-particle activation analysis.

Cpba : Competitive protein-binding assay.

Craa : Critical reflection activation analysis.

Critical absorption : In a pair of neighboring elements whose absorption edges for x-rays straddle the energy for a particular photon energy, the one that absorbs more strongly is called the critical absorber for the photon.

Critical mass : The minimum amount of fissile material necessary to sustain a chain reaction .

Critical reflection activation analysis : An activation analysis procedure for depth profiling of impurities in which activation is measured as a function of the angle of neutron reflection.

Cross bombardment : A method for assigning the mass to a radioactive nuclide by producing it in different nuclear reactions.

Cross fire : In autoradiography, the spread of images into film grains appearing over features other than those which emitted the particles.

Cross reaction : Ability of substances other than the analyte to bind to the binding reagent and ability of substances other than the binding reagent to bind the analyte in competitive binding assays.

Cross section : See cross section, microscopic .

Cross-linking : Covalent bonding between polymer chains.

Cross-section, activation : The cross section for the formation of a radionuclide by a specified reaction.

Cross-section, capture : The cross section for capture.

Cross-section, effective : In neutron induced reactions, an average cross section used to calculate a reaction rate per nucleus when the flux from the neutron source is known.

Cross-section, effective thermal : A fictitious cross section for a specified (neutron-induced) reaction which, when multiplied by the 2200 meter-per-second flux density , gives the correct reaction rate. Synonymous with Westcott cross section.

Cross-section, macroscopic : The cross section per unit volume of a given material for a specified process. For a pure nuclide , it is the product of the microscopic cross section and the number of target nuclei per unit volume; for a mixture of nuclides, it is the sum of such products.

Cross-section, partial : See cross-section, microscopic.

Cross-section, thermal : The microscopic cross section appropriate to thermal neutrons .

Cross-section, total : The sum of all partial cross sections.

Cross-section, westcott : See cross section, effective thermal.

Crud : Insolubly bulky corrosion products from fuel elements, condenser systems, water purification systems, etc., inside (water-cooled) nuclear reactors.

Cryptand : A molecular entity comprising a cyclic ligand assembly that contains three or more binding sites held together by covalent bonds and which defines a molecular cavity in such a way as to bind (and thus "hide" in the cavity) another molecular entity, the guest (a cation, anion, or a neutral species) more strongly than do the separate parts of the assembly (at the same total concentration of binding sites). The adduct thus formed is called a cryptate .

Cryptate : The adduct formed between a cryptand and a "guest" molecular entity, trapped within the cryptand's structure.

Crystal diffraction spectrometer : See spectrometer, crystal diffraction.

Cumulative fission yield : The fraction of fissions which have resulted in the production of a given nuclide , either directly or indirectly, up to a specified time. If no time is specified, the yield is considered to be the asymptotic value at scission .

Cumulative fraction released or **cumulative penetration** : A term for expressing leach rates of radionuclides from solidified waste forms based upon depletion of the radionuclide to a certain sample depth.

Curie : A unit of activity equal to exactly 3.7 X 1010 nuclear decays per second or 3.7 X 1010 becquerels.(Not capitalized. Not a SI unit. Symbol is "Ci".)

Curie (ci) : The original unit used to describe the intensity of radioactivity in a sample of material. One curie equals thirty-seven billion disintegrations per second, or approximately the radioactivity of one gram of radium. This unit is no longer recognized as part of the International System of units. It has been replaced by the becquerel.

Current density : Rate of flow across a surface per unit area of that surface.

Cutie pie : A colloquial term for a portable ionization chamber for determining relatively stable dose rates.

Cxms : Conversion X-ray Mssbauer spectroscopy.

Cyclic activation (analysis) : Activation analysis in which the sensitivity for short-lived radio activities is enhanced through accumulating spectra produced during repetitive short irradiations.

Cyclic instrumental neutron activation analysis : See cyclic activation analysis.

Cyclotron : Circular accelerator in which the particle is bent in traveling through a magnetic field, and an oscillating potential difference causes the particles to gain energy. ***OR*** A particle accelerator in which the particles travel in a succession of semicircular orbits of increasing radii under the influence of a magnetic field and are accelerated at the beginning of each such orbit by traversing an electric field produced by a high-frequency generator. These are sometimes classified as levels I, II, III, and IV, corresponding respectively to particle energies less than 10 MeV, 20 MeV, 45 MeV, and 200 MeV.

Cyclotron frequency : Frequency at which the electric field is switched in order to accelerate the particles in the cyclotron. The frequency is related to the mass and charge of the particle to be accelerated.

Dac : Derived air concentration.

Dalton : A unit of molecular weight used in some life sciences. Its use is discouraged.

Dating, radioactive : The determination of the radioactive age of an object from its content of radioactive substances and of their daughter products .

Daughter : A nuclide formed by the radioactive decay of a different (parent) nuclide.

Daughter (product) : Any nuclide which follows a specified radionuclide in a decay chain.

De broglie wavelength : In the quantum mechanical description of matter, the wavelength of a particle, from which all behavior is understandable, and whose magnitude relative to the environment determines whether or not classical physics is a justifiable approximation to reality.

Dead time : Constant and known value imposed on the resolving time by a paralysis circuit, usually in order to make the correction for resolving time losses more accurate.

Dead time correction : Correction to be applied to the observed number of pulses in order to take into account the number of pulses lost during the resolving or dead time.

Decay (nuclear) : A spontaneous nuclear transformation.

Decay (radioactive) : The change of one radioactive nuclide into a different nuclide by the spontaneous emission of radiation such as alpha, beta, or gamma rays, or by electron capture. The end product is a less energetic, more stable nucleus. Each decay process has a definite half-life.

Decay chain : A series of nuclides in which each member transforms into the next through nuclear decay until a stable nuclide has been formed. Synonymous with radioactive chain and radioactive decay series.

Decay constant : For a radionuclide, the probability α for the nuclear decay of one of its nuclei in unit time, i.e., the rate constant for radioactive decay, a first order reaction. It is given by $\alpha = (1/Nt)(dNt/dt)$, in which Nt is the number of nuclei of concern existing at time t. Synonymous with disintegration constant.

Decay curve : A graph showing the relative amount of radioactive substance remaining after any time interval.

Decay curve analysis : A mathematical procedure for deconvoluting a decay curve into the separate contributions from each radioactivity in a complex sample.

Decay energy : The total energy change when a nuclide in its ground state undergoes radioactive decay to a daughter product in its ground state. It is identical to the Q-value for the spontaneous reaction.

Decay rate : The ratio of activity to the number of radioactive atoms of a particular species.

Decay scheme : A graphical representation of the energy levels of the members of a decay chain showing the path by which nuclear decay occurs.

Decay series : See decay chain .

Decay time : The time required for a quantity to fall to ½ times the original value.

Decay, radioactive : Nuclear decay in which particles or gamma radiation are emitted or the nucleus undergoes spontaneous fission .

Decontaminant : Any chemical reagent(s) used in the act of decontamination .

Decontamination : The act of removing contamination from materials, organisms, an environment, etc.

Decontamination factor : The ratio of activity after decontamination to that before decontamination.

Dees : Regions of space in a cyclotron shielded from electric field in which the magnetic field causes the particles to bend in a semicircle.

Delayed beta decay : Radioactive decay to a daughter product which is unstable towards emission of a beta-particle which is then emitted in some fraction of decay events.

Delayed coincidence : A measurement using a coincidence circuit in which it is required that in order to be recorded, the occurrence of two events be separated by an electronically imposed time delay.

Delayed fission : Radioactive decay to a daughter product which is unstable towards fission and consequently undergoes fission in a measurable fraction of decay events.

Delayed neutron activation analysis : A method for measuring heavy elements, particularly uranium, thorium, and plutonium by activation to induce fission leading to delayed-neutron emitting products from which the neutrons are measured.

Delayed neutron emission : Radioactive decay to a daughter product which is unstable towards emitting a neutron and consequently does so in some fraction of decay events. Half-lives for delayed neutron emission are long compared to neutron emission associated directly with nuclear reactions or fission.

Delta rays : Secondary electrons, ejected by the primary ionizing interactions of charged particles passing through matter, provided their energy exceeds a few hundred eV.

Densitometry : Measurement of the response of an emulsion layer as blackening. The negative log of the transmittance of light is proportional to the electron exposure.

Density : The ratio of an object's mass to its volume.

Depleted material : Any substance in which the isotopic abundance of a stable isotope has been reduced from its natural value by artificial or natural means.

Derivative activation analysis : A method of activation analysis in which the element or chemical entity to be determined is either replaced by, or complexes with, a surrogate substance for which activation analysis has an enhanced sensitivity.

Derived air concentration : A concentration of a given radionuclide in air, obtained by means of a stylized model for the constantly maintained activity concentration of that radionuclide in air, which if breathed by the reference man for a working year of 2000 hours under conditions of light physical activity (breathing rate 1.2 m3/h) would result in an inhalation of one ALI (annual limit of air intake). Also the concentration which for 2000 hours of air immersion would lead to the irradiation of any organ or tissue to the appropriate limit.

Descendant : In a radioactive decay chain, any member of that chain except the parent.

Detection efficiency : The ratio between the number of particles or photons detected and the number of similar particles or photons emitted by the radiation source.

Detector : A neutron detector for which the cross section of the detection reaction varies inversely with neutron speed. *OR* A device or series of devices to used to measure nuclear particles and radiations.

Detector efficiency (, intrinsic) : The ratio of the number of particles or photons detected to the number of similar particles or photons which have struck face of a radiation detector. When multiplied by the geometry factor, it yields the absolute detector efficiency.

Detector, diffused junction semiconductor : A semiconductor detector in which the p-n or n-p junction is produced by diffusion of donor or acceptor impurities.

Detector, foil : See foil detector

Detector, liquid scintillator : A scintillation detector of which the scintillating medium is a liquid. The sample is often dissolved in the scintillating liquid.

Detector, radiation : An apparatus or substance for the conversion of radiation energy to a form of energy which is suitable for detection and or measurement.

Detector, scintillation : See scintillation detector.

Detector, semiconductor : See semiconductor detector.

Determinand : The element or chemical species to be determined.

Deuteron : A hydrogen nucleus of mass number 2.

Dichromatography : A technique based on the differential absorption of two x-rays or gamma-rays, which closely bracket the critical absorption edge of an element.

Diffused junction semiconductor detector : See semiconductor detector, diffused junction

Digestion : A chemical process for softening or solubilizing a material with heat, chemical reagents, and moisture.

Dis : Deep inelastic scattering.

Discriminator : A basic function unit comprising an electronic circuit which gives an output pulse for each input pulse whose amplitude lies above a given threshold value.

Disintegration constant : See decay constant .

Disintegration rate : See activity.

Disintegration, nuclear : Nuclear decay involving a splitting into more nuclei or the emission of particles.

Displacement analysis : One of several general terms for competitive binding assay , most often used to refer to a variant technique wherein the addition of tracer is delayed to enhance assay sensitivity.

Displacement, radioactive : Originally established empirically, this connects the type of radioactive decay (α) with the displacement, caused by the change in nuclear charge, of the daughter relative to the parent in the periodic table of elements.

Distribution ratio : The ratio of the total analytical concentration of a substance in the stationary phase to its total analytical concentration in the mobile phase, usually measured at equilibrium.

Dna : deoxyribonucleic acid

Donor : In a solid state detector, an impurity that is added to a pure semiconductor material to increase the number of free electrons. M.

Doppler broadening : Frequency spreading that occurs in single-frequency radiation when the radiating atoms do not all have the same velocity and may each give rise to a different Doppler shift .

Doppler shift : The amount of change in the observed frequency of a wave due to the Doppler effect, that is, relative motion of the source and detector.

Doppler shift attenuation : The dependence of a gamma-ray line shape on by velocity changes in a decelerating ion. The effect can be used to determine the stopping power of heavy ions in various materials if an emitter with a known lifetime is used.

Dose : A general term denoting the effect of absorption of a quantity of radiation or energy absorbed. ***OR*** A general term denoting the quantity of radiation (energy) absorbed. For special purposes, it must be appropriately qualified, c. q. absorbed, maximum permissible, mean lethal.

Dose equivalent (, effective) : The absorbed dose multiplied by the quality factor and the product of all other modifying factors N, aimed at expressing on a common scale, for different types of radiations and distributions of absorbed dose , the biological effects associated with an exposure. Also called just the dose equivalent.

Dose meters or **dosimeters** : An instrument that measures the total dose of nuclear radiation received during a given time period.

Dose, absorbed : The energy imparted to matter by ionizing radiation in a suitable small element of volume divided by the mass of that element of volume.

Dosimetry : The technique of measurement of dose .

Double irradiation technique : In activation analysis, the irradiation of identical samples at positions of very different fast-to-thermal neutron flux. The technique is used to resolve interferences of different analyte elements leading to the same final product.

Double labeling : Labeling with two isotopes at two different sites on the same molecule.

Double radioimmunoassay : A radioimmunoassay in which small amounts of antibody-antigen complex serve as antigen in a second antibody-antigen reaction that acts to scavenge the labeled substance.

Doubling time : The amount of time required before the quantity of a substance doubles when the rate of growth is approximately exponential.

Drift (of a charged particle) : The movement of current carriers in a semiconductor under the influence of an applied voltage.

Dscems : Depth selective conversion electron Mssbauer spectroscopy.

DTPA : Diethylenetriamine pentaacetic acid (pentetic acid)

Dvi : Prefix to an element indicating another element in the same column but two rows lower in the periodic table having the light elements located at the top. Derived from the Sanskrit word for two.

Dwell time : The time increment during which pulses are stored in a multichannel pulse height analyzer operating in the multiscaling mode. (See multiscaling.)

Dynodes : An electrode, used in photomultipliers, whose primary function is secondary emission of electrons.

Ebq : Exabecquerel (10^{18} Bq)

Ec : Electron capture (q. v.)

Ecd : See electron capture detector .

Edta : A metal chelating agent, ethylene-diaminetetraacetic acid.

Edta : Ehylenediaminetetraacetic acid (edetic acid).

Edxrf : See energy dispersive x-ray fluorescence analysis.

Effective cadmium cut-off (energy) : In a given experimental configuration, the energy value determined by the condition that the detector response would be unchanged if the cadmium cover surrounding the detector was replaced by a fictitious cover opaque to neutrons with energy below this value and transparent to neutrons with energy above this value.

Effective thermal cross-section : See cross section, effective thermal.

Efficiency (of a counter) : See counting efficiency.

Efficiency, intrinsic : See detector efficiency, intrinsic.

Effluent, radioactive : Any solid, liquid or gaseous radioactive waste material discharged from a system.

Eka : Prefix to an element indicating another element in the same column but one row lower in the periodic table having the light elements located at the top. Derived from the Sanskrit word for "one". This nomenclature scheme has been used historically for designating new, or unknown elements. Thus, gallium was eka-aluminum; element 118 would be eka-radon.

Elastic recoil detection analysis : A type of ion beam analysis and Rutherford backscattering spectroscopy in which an ion beam strikes a sample at a grazing angle and recoiling ions are detected. The technique is used for depth profiling, particularly for light-elements in a heavy matrix.

Elastic scattering : See scattering, elastic.

Electric dipole moment : The product of charge and distance of separation for an electric dipole.

Electrodeposition : A method of thin sample preparation in which an electric current is used to deposit sample material.

Electromagnetic radiation : Electromagnetic waves and the associated energy; photons.

Electromagnetic radiation : Radiation consisting of electric and magnetic fields that travels at the speed of light. Examples light, radio waves, gamma rays, x-rays.

Electrometer : A device for measuring electric charges or electric currents.

Electron : An elementary particle with a unit electrical charge and a mass 1/1837 that of the proton. Electrons surround an atom's positively charged nucleus and determine that atom's chemical properties. ***OR*** A stable elementary particle having an electric charge of 1.60219 X 10-19 C and a rest mass of 9.1095 X 10-31 kg. When used without specification the term means the negatively charged electron, which is also called the negatron. Its anti-particle, the positively charged electron, is called positron .

Electron accelerator : An accelerator in which the beam particles are electrons .

Electron capture : A radioactive decay process in which an orbital electron is captured by and merges with the nucleus. The mass number is unchanged, but the atomic number is decreased by one.

Electron capture detector : 1) A very sensitive detector for the presence of certain classes of molecules, most notably chlorine-containing molecules. It is based on a steady current produced by a long-lived beta-emitter and the measurable perturbations in that current caused by the presence of small quantities of molecules that scavenge or capture the electrons. 2) A detector for electron capture decay.

Electron spectroscopy for chemical analysis : Type of spectroscopy involving the measurement of the kinetic energy of electrons emitted by chemical substances usually as a result of excitation by monochromatic X-rays.

Electron volt : A unit of energy corresponding to 1.6022X10-19 J.

Electron, aqueous : A hydrated electron; an electron solvated by water molecules.

Electron, compton : See Compton electron.

Electron, conversion : See conversion electron.

Electron, secondary : An electron emitted as a result of momentum transfer from some primary radiation.

Electron-volt (ev) : Energy unit used as the basis of measurement for atomic (eV), electronic (keV), nuclear (MeV), and subnuclear processes (GeV or TeV). One electron-volt is equal to the amount of energy gained by an electron dropping through a potential difference of one volt, which is 1.6×10^{-19} joules.

Electroscope : A device for measuring very small electrostatic charges.

Electrospraying : A soft (non-violent) sample ionization technique, applicable particularly to large and complex species in solution, for producing intact ions in vacuo. Solutions pass through a needle maintained at a few kilovolts relative to a chamber. The resulting field at the needle tip charges the surface of the emerging liquid, dispersing it by Coulomb forces into a spray of small charged droplets.

Electroweak interaction : A theory which unifies the electromagnetic and weak interactions.

Element : A substance in which the nucleus of each atom is characterized by having exactly the same number of protons.

Elementary particle : See particle, elementary.

Eluant : Liquid used to extract one material from another as in chromatography. The term eluant is preferred. *OR* The effluent emerging from a chromatographic bed when elution is carried out. *OR* The liquid or gas entering a chromatographic bed and used to effect a separation by elution.

Elute : To remove or separate by elution chromatography.

Emanation : 1. Radon 2. Any radioactive gas released by another radioactive substance.

Emanation power : A ratio of the amount of radon escaping a solid or solution, to the amount being formed.

Emanation thermal power : A thermoanalytical technique in which the release of radioactive emanation from a substance (and/or its reaction products) is measured as a function of temperature while the substance is subjected to a controlled temperature program.

Emanometric analysis : A method of quantitative analysis which uses measurement of the radioactive isotopes of inert gases for the determination of appropriate elements.

Emp : electro-magnetic pulse

Emulsion, nuclear : An ionization-sensitive material such as a photographic emulsion used for permanently recording the tracks of charged particles.

Enaa : Epithermal neutron activation analysis.

End labeling : Procedure in which terminal phosphate groups of DNA are removed and enzymatically replaced with radiolabeled groups.

End product : The final product of a nuclear reaction or process.

Energy (of a radiation) : Energy of the individual particles or photons of which radiation consists.

Energy dispersive x-ray fluorescence analysis : A method of X-ray fluorescence analysis where element specificity is obtained by measuring the energy spectrum of the induced X-radiation.

Energy flux density : See flux density, energy.

Energy level : The stable energy state that an atom or molecule or nucleus can assume.

Energy resolution : A measure, at given energy, of the smallest difference between the energies of two particles or photons capable of being distinguished by a radiation spectrometer. It is usually expressed as FWHM. See full-width at half-maximum.

Energy threshold : The limiting kinetic energy of an incident particle or energy of an incident photon (both expressed in the laboratory system) below which a specified process cannot take place.

Energy, binding : See binding energy.

Enrichment : Any process by which the isotopic abundance of a specified isotope in a mixture of isotopes of an element is increased.

Enrichment factor : For a material enriched in a specified isotope , the ratio between the isotopic abundance and the natural abundance of that isotope.

Enrichment, isotopic : Any process by which the isotopic abundance of a specified isotope in a mixture of isotopes of an element is increased. C.

Eob : End-of-bombardment.

Epicadmium neutrons : See neutrons, epicadmium .

Epithermal neutron activation analysis : Activation analysis using epithermal neutrons, i. e., those with energies usually just above thermal.

Epithermal neutrons : See neutrons, epithermal.

Epsilon rays : Tertiary electrons. See delta rays.

Equilibrium isotope effect : See isotope effect.

Equilibrium, radioactive : Among the members of a decay chain, the state which prevails when the ratios between the activities of successive members remain constant.

Equilibrium, secular : Radioactive equilibrium where the half life of the precursor isotope is so long that the change of its activity can be ignored during the period of interest and all activities remain constant.

Equilibrium, transient : Radioactive equilibrium where the ratio of the (longer-lived) parent to (shorter-lived) daughter activity as a function of time is a constant less than unity.

Er : Enhanced radiation

ERD or **ERDA** : Elastic recoil detection analysis.

Esca : Electron spectroscopy for chemical analysis.

Escape peak, double : In a gamma-ray spectrum, the peak due to pair production in the detector and escape, from the sensitive part of the detector, of the two photons of 511 keV resulting from annihilation.

Escape peak, iodine : The x-ray escape peak in a gamma-ray spectrum using a NaI scintillation spectrometer and due to the escape of the 28 keV X-ray of the iodine in the scintillation crystal.

Escape peak, single : In a gamma-ray spectrum, the peak due to pair production in the detector and escape, from the sensitive part of the detector, of one of the photons of 511 keV resulting from annihilation.

Escape peak, x-ray : In a gamma or X-ray spectrum, the peak due to the photoelectric effect in the detector and escape, from the sensitive part of the detector, of the X-ray photon) emitted as a result of the photoelectric effect.

Eti : Early transient incapacitation

Eti-pd : Early transient incapacitation and performance decrement

Ev : Electron volt

Exchange rate : The rate at which an exchange reaction occurs.

Exchange reactions : Reactions in which two atoms or ions exchange places, either in two molecules or in the same molecule.

Excitation : Process causing the transition of a system from one state to another of higher energy.

Excitation energy : Minimum energy required to bring a system to a specified higher energy level.

Excitation function : In nuclear reactions the dependence of the cross section for the reaction upon the energy of the projectile .

Excited state : State of a system with energy higher than that of the ground state .

Excited state : The state of an atom or nucleus when it possesses more than its normal energy. Typically, the excess energy is released as a gamma ray.

Exponential decay : Variation of a quantity (generally the activity of a radionuclide) according to the law A=Aoe-t where A and Ao are the values of the quantity being considered at time t and zero respectively, and ? is an appropriate constant.

Exposure : For X- or gamma radiation in air: the sum of the electrical charges of all the ions of one sign produced after all electrons liberated by photons in a suitably small volume element of air are completely stopped, divided by the mass of the air in the volume element.

Extractable species : Any chemical species which can be separated through an extraction procedure.

Extraction : A separation method in which a liquid solvent causes the transfer of one or more analytes into it from contact with a second liquid or solid phase mixture.

Extraction chromatography : Synonymous with liquid-liquid chromatography. One liquid can be a film adsorbed on a solid stationary packing material.

Extraction cycle : A series of steps involving solvent extraction, stripping, and in some cases scrubbing .

Extrapolated range : The distance from a radiation source at which the flux density has decreased to one-half of its initial value. It is calculated by extrapolation of the tangent to the flux density) versus distance curve to zero flux density.

Fallout, radioactive : The deposition on the ground of radioactive substances from nuclear explosions and other injections of radioactive material into the atmosphere.

Families, radioactive : The naturally occurring decay chains associated with uranium-238, uranium-235, and thorium-232, and the artificial decay chain associated with neptunium-235.

Faraday cup : A hollow collector, open at one end and closed at the other, used to collect and measure, beams of ions.

Fast : Fast or forward alpha scattering technique .

Fast neutron activation analysis : Activation analysis using fast neutrons , usually from accelerator sources, in the MeV energy range, most frequently 14 MeV.

Fast neutrons : See neutrons, fast .

Feather analysis : A technique for determining the range of a nuclide's beta rays in aluminum by comparing the absorption curve to that of a reference beta-emitting species.

Feather rule : A rule or equation for the maximum range of beta particles as a function of particle maximum energy. The equation is usually a linear one and when expressed in terms of mass per unit area does not vary widely with the atomic number of the absorber.

Fena : Filtered epithermal neutron activation.

Fermi : A unit of length useful on the nuclear scale and defined as 1 fermi = 1 fm = 10-15 meters. (Not a SI unit.) Synonymous with femtometer.

Fermion : A particle having a spin that is an odd integer multiple of (h-bar)/2.

Fertile : 1. Of a nuclide : capable of being transformed, directly or indirectly, into a fissile nuclide by neutron capture . 2. Of a material: containing one or more fertile nuclides.

Fet amplifier : A field-effect transistor amplifier.

Figure of merit : A measure of performance of a detection system that combines sensitivity and ability to record rapid or discontinuous spatial changes in source intensity.

Film badge : A package containing one or more small photographic films for approximate measurement of radiation exposure, used for the purpose of monitoring personnel.

Filter (of a radiation) : Material interposed in the path of heterogeneous radiation to modify the spectral distribution of the radiation.

Filtered epithermal neutron activation analysis : Epithermal neutron activation analysis in which various filters are used, frequently to change the influence of resonances on reaction probabilities.

Fingerprint, nuclear : A pattern of distinct radiation energies and type suitable to identify the emitting substance.

Fissile : 1. Of a nuclide : capable of undergoing fission by interaction with (slow) neutrons. 2. Of a material: containing one or more fissile nuclides .

Fissile nucleus : A nucleus that may fission after collision with a thermal (slow) neutron or that fissions spontaneously (by itself).

Fission : The splitting of a heavy nucleus into two roughly equal parts (which are nuclei of lower-mass elements), accompanied by the release of a relatively large amount of energy in the form of kinetic energy of the two parts and in the form of emission of neutrons and gamma rays. *OR* Spontaneous or induced Nuclear transformation in which a nucleus splits into two or more relatively large fragments and can be referred to in more detail as binary, ternary,... fission.

Fission chamber or **counter** : An ionization chamber , used to detect slow neutrons, the inside of which is coated with uranium from which highly ionizing, recoiling fission products may be produced.

Fission fragments : Nuclei resulting from fission and possessing kinetic energy acquired from the fission process.

Fission isomer : A highly deformed metastable nuclear state of a fissile nuclide.

Fission neutrons : See neutrons, fission .

Fission products : Nuclides produced by fission and the daughter products of these nuclides. OR Nuclei formed by the fission of higher mass elements. They are of medium atomic mass and almost all are radioactive. Examples ^{90}Sr, ^{137}Ce.

Fission yield : The fraction of fissions giving rise to a particular nuclide before any nuclear decay has occurred.

Fission yield, chain : The fraction of fissions giving rise to nuclei of particular mass number.

Fission yield, cumulative : The fraction of fissions which have resulted in the production of a nuclide either directly or indirectly, up to a specified time. If no time is specified, the yield is considered to be the asymptotic value.

Fission yield, direct : The fraction of fissions giving rise to a particular nuclide before any nuclear decay has occurred.

Fission yield, independent : The fraction of fissions which have resulted in the direct production of a given fission product nuclide .

Fission, nuclear : The spontaneous or induced division of a nucleus into two or more major parts, usually accompanied by the emission of neutrons, gamma radiation and, rarely, small charged nuclear fragments.

Fission, spontaneous : Nuclear fission which occurs without the addition of particles or energy to the nucleus.

Fission, thermal : Fission caused by thermal neutron.

Fissionable : See fissile .

Fixation : The incorporation of radioactive elements, usually fission products, into solid materials in such a way as to insure no significant release over long periods of exposure to the natural environment.

Flow counter : A radiation counter in which an appropriate atmosphere is maintained in a counter tube by allowing a suitable gas to flow through the volume. Such counters are operated in either the Geiger-Muller region or the proportional region.

Fluence : A measure of the time-integrated particle flux expressed in particles per unit area.

Fluor : A liquid or solid that is used in scintillation counters and that emits a flash of light when it is excited by radioactive or other radiation.

Fluorescence : Luminescence which occurs essentially only during the irradiation of a substance by electromagnetic radiation.

Fluorescence yield : For a given excited state of a specified atom , the ratio of the number of excited atoms which emit a photon to the total number of excited states.

Fluorimetry : A method of analysis involving fluorescence by an analyte.

Fluorography : Autoradiography involving the additional use of scintillation fluid or solid. Jones. Also known as photofluorography.

Fluorometry : See fluorimetry.

Flux : The amount of a substance passing through an area per unit time. See also flux density.

Flux density : In a multicomponent mixture, the flux density NB of a species B is a vector which indicates the direction in which the species moves and the amount of substance of B passing through a plane perpendicular to the vector, divided by the time and by area. Sometimes the flux density is simply called flux . However, this is not consistent with the usage in vector theory.

Flux density, 2200-metre-per-second : A fictitious flux density expressed as the product of the total number of neutrons per cubic meter and a neutron speed of 2200 meters per second.

Flux density, energy : For mono-directional radiation, the energy traversing in a time interval a small area perpendicular to the direction of the energy flow, divided by that time interval and by that area.

Flux density, neutron : See flux density, particle .

Flux density, particle (or, photon) : At a given point in space, the number of particles or photons incident in a time interval on a suitably small sphere centered at that point, divided by the cross-sectional area of that sphere and by that time interval. The particle flux density is identical with the product of the particle density and the average speed of the particles.

Flux depression : The lowering of the particle flux density in the neighborhood of a sample due to absorption of particles in the sample.

Flux monitor : A known amount of activatable material irradiated together with a sample; the induced radioactivity is used as a measure for a particular flux density during the irradiation.

Flux perturbation : The lowering of the flux density in a sample as a result of both flux depression and self shielding . Sometimes a specified flux density can increase as a result of the change of the energy of the particles in the sample, e.g., the increase of the thermal neutron flux density in hydrogen containing material.

Fnaa : Fast neutron activation analysis.

Foil detector : A small piece of foil used to measure flux densities by activation.

Force, nuclear : The force acting between nucleons. Synonymous with strong force.

Forward alpha scattering technique : Determination of elastically scattered alpha particles with a detector at forward angles; the method uses the energy of the detected particle to deduce the scattering mass for light target elements.

Four- counting : Measuring radiation from a source with a detector or detector system that covers all directions from the source; that is, with a 100% geometry factor .

Fraction, bound : In radioassays, the fraction of the incubation mixture which, after separation, contains the analyte bound to the binding reagent.

Fraction, free : In radioassays, the fraction of the incubation mixture which, after separation, does not contain the bound analyte.

Fractional crystallization : A class of chemical separation by fractionation which takes advantage of slight differences in solubility of different compounds in a solvent, the least soluble crystallizing first and removed from the mother liquor .

Free radical : See radical .

Frenkel defect : A crystal defect consisting of a vacancy and an interstitial atom which arises when an atom is moved out of a normal lattice site and forced into an interstitial position.

Frequency of emission : In imaging with gamma rays, it is the branching fraction of each gamma ray.

Fricke dosimeter : A dosimeter based on the oxidation of aqueous Fe^{+2} to Fe^{+3} by ionizing radiation.

Fuel cycle : The sequence of steps, such as mining, processing, fabrication, utilization, reprocessing, and refabrication, through which nuclear fuel may pass.

Fuel element : The smallest structurally discrete part of a reactor which has fuel as its principal constituent.

Fuel reprocessing : The processing of nuclear fuel , after its use in a reactor, to remove fission products and recover fissile and fertile material.

Fuel, nuclear : Material containing fissile nuclides , which when placed in a reactor enables a chain reaction to be achieved.

Full energy peak : In a radiation spectrum , the part of the spectral response curve corresponding to the total energy of the detected radiation .

Full energy peak efficiency, absolute : For a radiation spectrometer viewing a source of radiation, the fraction of events of a given energy radiation that are registered in that radiation's full energy peak .

Full energy peak efficiency, intrinsic : The detector efficiency when considering only events where the total energy of the

		radiation is absorbed in the sensitive volume of the detector.
Full width at half maximum (fwhm)	:	In a distribution curve comprising a single peak, the width of that peak at half its maximum amplitude.
Full width at tenth maximum (fwtm)	:	In a distribution curve comprising a single peak, the width of that peak at one-tenth its maximum amplitude.
Fusion	:	A process whereby low mass nuclei combine to form a more massive nucleus plus one or move massive particles.
Fusion reaction, nuclear	:	A reaction between any two nuclei which results in their complete coalescence to give a compound nucleus equal to the sum of the two constituent nuclides.
Fusion, nuclear	:	The process in which nuclei undergo nuclear fusion reactions.
Fwhm	:	See full width at half maximum.
Fwtm	:	See full width at tenth maximum.
G	:	See G-value .
Gain (multiplier tube)	:	Ratio of output to input electrical power.
Gamma cascade gamma quantum	:	Two or more different gamma rays emitted successively, that is, in coincidence , as a single nucleus passes through one or more energy levels.
Gamma radiation	:	Electromagnetic radiation emitted in the process of nuclear transformation or particle annihilation.
Gamma radiation, capture	:	See capture gamma radiation.
Gamma ray	:	A highly penetrating type of nuclear radiation, similar to x-radiation, except that it comes from within the nucleus of an atom, and, in general, has a shorter wavelength.
Gamma-ray spectrometer	:	A system for determining the energy spectrum of gamma rays .

Gamma-ray spectrometer, anti-compton : A gamma-ray spectrometer in which the effect of the Compton scattering is at least partly compensated by an anticoincidence requirement on the full energy peak .

Gamma-ray spectrum : A spectrum of photon radiation.

Gamow barrier : The potential barrier which retards the escape of alpha-particles from the nucleus.

Gas jet : A gas transport system using a stream of helium gas to thermalize and very rapidly transport recoil products from nuclear reactions to a collector and detection apparatus.

Gas-flow counter : A proportional counter tube in which the gas between the cathode and anode is constantly, but slowly, replenished through flow.

Gas-transport system : See gas jet .

Gauge boson : Particle mediating an interaction. By exchange of the gauge particle, the interaction between two particles is accomplished.

Geiger counter : A Geiger-Muller detector and measuring instrument. It consists of a gas-filled tube that discharges electrically when ionizing radiation passes through it and a device that records the events.

Geiger-muller counter tube : See counter tube, Geiger-Muller.

Geiger-muller region : The range of operating voltage of a counter tube in which each ionizing event gives rise to an output pulse having an amplitude independent of the number of ions initially produced in the sensitive volume by that ionizing event.

Geiger-muller threshold : The minimum voltage necessary for a counter tube to operate in the Geiger-Muller region.

Generator : See radioisotope generator.

Genetic relation : A reference to the connection between two nuclides which are members of the same decay chain .

Geometric attenuation : The reduction of radiation due only to the effect of the distance between the point of interest and the source, excluding the effect of any matter present.

Geometry (,counting) : A term used colloquially to signify the arrangement in space of the various components in an experiment, particularly the source and the detector in radiation measurements.

Geometry factor : The average solid angle in steradians at the source subtended by the aperture or sensitive volume of the detector , divided by 4.

Germanium detector : A diffused junction semiconductor detector in which germanium is the base solid state material.

Germanium detector, ge, Germanium detector, high purity : A diffused junction semiconductor detector into which lithium has been diffused from the surface and which must be maintained at low (liquid nitrogen) temperatures. It is used in high resolution gamma ray spectrometry.

Germanium detector, intrinsic : See germanium detector, high purity.

Glove box : An enclosure in which material may be manipulated in isolation from the operator's environment. This manipulation is effected by gauntlet gloves or flexible plastic devices fixed to ports in the walls of the box.

Gluon : A gauge particle mediating the color strong interaction.

Gm counter : See counter tube, Geiger-Muller.

Grain yield : The average number of silver grains per electron strike on the recording layer of autoradiography film.

Granddaughter : In a sequence of radioactive decays , the daughter of the daughter.

Gray : SI unit of absorbed dose ; 1 gray = 1 joule/ kilogram.

Gray (gy) : Unit of absorbed dose due to any type of radiation. An exposure to 1 gray results from radiation depositing one joule per kilogram of animal tissue or any other material.

Grenz rays : X-rays produced in the region of 5 to 20 kV.

Ground state : The state of lowest energy of a system.

Group preconcentration : An operation (process) as a result of which several components are isolated in one step. This may be achieved in one of two ways: by transport either of the matrix or of the microcomponents into a second phase.

Group separation : See group preconcentration .

Growth and decay curve : A graph showing the relative amount of radioactivity remaining after any time interval and in which the radioactivity also shows growth from a parent nuclide.

Growth curve (of activity) : Curve giving the activity of a radio-active nuclide as a function of time and showing the increase of the activity through the decay of the parent substance or as a result of activation.

Guard foil : In an irradiation of a target, a foil that is used on one side of that target to exactly compensate for nuclear reaction products that recoil out of the opposite side and are otherwise lost.

G-value : The number of specified chemical events produced in an irradiated substance per 100 eV of energy absorbed from ionizing radiation .

Gy : Symbol for the absorbed dose unit gray . Not to be confused with an old notation for gigayear, also Gy, but now Ga

Gy : Gray

Hadron : A strongly interacting particle.

Half life, biological : For a substance, the time required for the amount of that substance in a biological system to be reduced to one-half of its value by biological processes, when the rate of removal is approximately exponential.

Half life, effective : For a radioactive substance, the time required for the amount of that substance in a biological system to be reduced to one half of its value by both radioactive decay and biological processes, when the rate of removal is approximately exponential.

Half life, partial : In nuclei undergoing branching decay , the half-life that would be evident if a particular branch were the sole means of decay. The partial half-life of a nucleus by a specific mode of decay is the (total) half-life divided by the branching fraction for that specific decay path.

Half life, radioactive : For a single radioactive decay process, the time required for the number of atoms or the activity to decrease to half its value by that process.

Half thickness : The thickness of a specified substance which, when introduced into the path of a given beam of radiation , reduces the value of a specified radiation quantity by one half.

Half-life : The time in which half the (large number of) atoms of a particular radioactive nuclide disintegrate. The half-life is a characteristic property of each radioactive isotope.

Half-value layer/Half-value thickness : See half thickness.

Halo : A luminous or colored circle seen in the light around an object; frequently applied to radioactive inclusions in minerals.

Hap : Hydrated antimony pentoxide, an inorganic ion-exchanger.

Hapten : A substance, particularly of low molecular weight, which cannot to induce antibody formation by itself, but which when combined with larger molecules such as proteins can do so. This combination is called an immunogen .

Hard radiation : Radiation whose particles or photons have a high energy and, as a result, penetrate materials relatively easily.

Hardening (neutrons) : The increase in the fraction of higher energy neutrons in a reactor beam flux due to preferential absorption of low-energy neutrons by a sample.

Hd value : In autoradiography , a measure of resolution corresponding to the distance from a straight radiographic line that accounts for 50% of the grains.

Health physics : A division of occupational health dealing particularly with protection of personnel from the harmful effects of ionizing radiation .

Heavy elements : A term often used with reference to the actinides and still heavier elements.

Heavy ion : A nucleus consisting of more than one nucleon. Frequently, the term is further restricted to nuclei with atomic numbers greater than two.

Heavy ion accelerator : An accelerator in which the beam consists of heavy ions.

Heavy ion reaction : A nuclear reaction in which the interacting nuclei are heavy ions .

Heavy water : Water containing a significant fraction (up to 100%) of deuterium in the form of D2O or HDO.

Helium jet : See gas jet .

Heterogeneous counting : A technique in solid state scintillation counting of radiochromatograms in which the scintillant is solid bead material with which a flow cell is packed and that does not chemically interact with the eluting medium.

Hfr : High flux reactor.

Hiaa : Heavy-ion activation analysis.

High flux reactor : A nuclear reactor designed to produce neutron flux densities several orders of magnitude greater than in conventional reactors.

Hmd : Hydrated manganese dioxide, an inorganic ion-exchanger.

Hob : height-of-burst

Hold-back carrier : See carrier, hold back

Homogeneous counting : A technique for counting radio-chromatograms in a flow cell by injecting liquid scintillant into the eluting medium. Jones.

Homolog (or homologs) : Elements in the same periodic table group that tend to exhibit similar, but not identical, chemical properties.

Hormesis : Controversial theory which argues that there is a benefit to health, or decrease in biological damage from radiation as dose in increased (valid only for very small doses).

Hot atom : An atom in an excited energy state or having kinetic energy above the ambient thermal level, usually as a result of nuclear processes.

Hot cell : A heavily shielded enclosure for highly radioactive materials. It may be used for their handling or processing by remote means or for their storage.

Hot laboratory : A heavily shielded, specially equipped laboratory to protect personnel working with large amounts of substances emitting gamma radiation and to protect the environment from contamination.

Hpge : See germanium detector, high purity . Also known as intrinsic germanium.

Hplc : High performance liquid chromatography.

Hubble constant : Ratio of outward speed of galaxies to their distances from Earth.

Hvl : half-value layer

Hwr : Heavy water reactor, that is, one moderated and cooled with heavy water .

Hydrolysis : Reaction with water as a solvent involving the rupture of one or more bonds in the reacting solute. For weak electrolytes, this can alter the acidity and subsequent chemical behavior of the aqueous solvent.

Hyperfine structure : Small differences in energy states of atoms or molecules that have different orientations between nuclear spin and electron angular momentum owing to the interaction between the nuclear and electron magnetic moments.

Iba : Ion beam analysis .

Icpms : Inductively coupled plasma mass spectrometry.

Ida : Isotope dilution analysis .

Idms : Isotope dilution mass spectrometry.

Idms : Isotope dilution mass spectrometry.

Immunoassay : An assay procedure based on the reversible and non-covalent binding of an antigen by an antibody . Immunoassays can be employed to detect or quantify either antigens (haptens) or antibodies.

Immunoassay : An assay procedure based on the reversible and non-covalent binding of an antigen by an antibody . Immunoassays can be employed to detect or quantify either antigens (haptens) or antibodies.

Immunogen : A substance that induces an immune response, one that stimulates the production of specific antibody or lymphocytes.

Immunoradiometric assay : An assay procedure based on the reversible and non-covalent binding of an antigen by a specific antibody labeled with a radioactive nuclide as tracer. Synonymous with radiometric assay.

Immunoradiometric assay, two-site : An immunoradiometric assay involving two sets of antibodies, one of which is labeled, that combine with different immunoreactive sites of an antigen molecule.

In source : See isotopic neutron source .

Inaa : Instrumental neutron activation analysis .

Independent fission yield : See fission yield, independent.

Indicator, radioactive : A radioactive substance which functions as an indicator for various chemical, physicochemical, etc. processes, e.g., absorption, precipitation or extraction. *OR* Radioactivity induced by irradiation. *OR* Radioactivity that is created by bombarding a substance with neutrons in a reactor or with charged particles produced by particle accelerators.

Inelastic scattering : Scattering in which the total kinetic energy of the two collision partners changes.

Infinite source thickness : For a specified radiation , the minimum thickness of a flat preparation of a radioactive material where the intensity of the specified radiation at the surface does not increase when the thickness is increased by adding more of similar radioactive material.

Infinite source thickness : For a specified radiation , the minimum thickness of a flat preparation of a radioactive material where the intensity of the specified radiation at the surface does not increase when the thickness is increased by adding more of similar radioactive material.

Infrared radiation : Electromagnetic radiation of longer wavelength than visible light.

Inner bremsstrahlung : See bremsstrahlung, inner.

Instrumental activation analysis : See activation analysis, instrumental.

Instrumental photon activation analysis : Activation analysis in which high-energy photonuclear reactions are used for the activation.

Integral flux : The total particle flux density encompassing all particle energies.

Integrated flux : See fluence.

Intensity of radiation : The energy per unit time entering a small sphere in a time interval, divided by the area of the great circle of the sphere and the time interval. For mono-directional radiation , this reduces to the energy flux density.

Intercomparison, analytical : A procedure which gives insight into the accuracy of results of analytical procedures by comparing the results obtained in the analyses of identical samples at different laboratories and preferably with different analytical methods.

Interference, nuclear : In activation analysis for a particular element producing a radionuclide, the formation of that same radionuclide from another element.

Interference, spectral : An interference caused by 1.) unresolved overlapping with a neighboring spectral peak, 2.) disturbing of proper spectrometer response at a peak due to high levels of background or of other radionuclides.

Internal conversion : See conversion, internal.

Internal conversion coefficient : See conversion coefficient, internal.

Internal conversion coefficient : See conversion coefficient, internal.

Internal monitor : Beam monitoring in which the monitor is part of the target or target assembly being irradiated.

Interstitial : Pertaining to or situated in the interspaces of a tissue or other structure.

Intrinsic efficiency : See detector efficiency .

Intrinsic full energy peak efficiency : See full energy peak efficiency, intrinsic .

Intrinsic ge detectors : See Germanium detector, high purity .

Intrinsic photopeak efficiency : See photopeak efficiency, intrinsic .

Inverse square law : The intensity of radiation from a point source in free space is inversely proportional to the square of the distance from the source. Presence of absorbers necessitates corrections to this law.

Ion : An atomic or molecular particle having a net electric charge.

Ion beam : A charged-particle beam.

Ion beam analysis : A collective term for any of a variety of techniques that involve irradiation of a sample with an ion beam for the purpose of analysis. Included are particle-induced gamma and x-ray emission, Rutherford backscattering spectroscopy, nuclear microprobes and other methods of nuclear reaction analysis .

Ion density : The number of ions per unit volume. Also known as ion concentration.

Ion exchange : 1. In surface chemistry, if the adsorption of one or several ionic species is accompanied by the simultaneous desorption (displacement) of an equivalent amount of one or more other ionic species, this process is called ion exchange. 2. The process of exchanging ions between a solution and an ion exchanger .

Ion exchange chromatography : Separation based mainly on differences in the ion-exchange affinities of the components.

Ion exchanger : A solid or liquid, inorganic or organic, containing ions exchangeable with others of the same sign present in a solution in which the exchanger is considered to be insoluble. A monofunctional ion exchanger contains only one type of ionogenic group, that is, a group which is either ionized or capable of dissociation into fixed ions and mobile counter-ions.

Ion geminate pairs : Ion pairs in which oppositely charged particles are held together at certain equilibrium distance by Coulomb attraction.

Ion microprobe : The probe used in ion probe microanalysis.

Ion pair : A pair of oppositely charged ions, or an electron and a cation, held together by Coulombic attraction without formation of a covalent bond.

Ion probe microanalysis : Any technique in which the specimen is bombarded by a focused beam of (primary) ions (diameter less than 10 μm) and the (secondary) ions ejected from the specimen are detected after passage through a mass spectrometer .

Ionization : The formation of ions by addition or removal of electrons from atoms , molecules, or groups of molecules or by the division of molecules.

Ionization chamber : A radiation detector which employs an electric field for the collection at the electrodes of charges associated with the ions produced in the sensitive volume by ionizing radiation , without charge multiplication. The solid angle (2, 4) subtended when using the detector is sometimes specified.

Ionization current : The passage of a charged particle through a gas causes ionization; if an electric field is applied across the gas a current (due to the motion of the ions produced) flows between the electrodes.

Ionization density : The density of the ionization, produced in the path of a charged particle passing through matter, depends upon the amount of charge, the velocity of the particle, and the nature of the matter.

Ionization energy : The minimum energy required to ionize an atom or molecule which is originally in the ground state.

Ionization potential : See ionization energy .

Ionization, specific : See specific ionization .

Ionizing radiation : Radiation that is capable of producing ions either directly or indirectly. ***OR*** Any radiation consisting of directly or indirectly ionizing particles or a mixture of both or photons with energy higher than the energy of photons of ultraviolet light or a mixture of both such particles and photons.

Ipaa : See instrumental photon activation analysis.

Ipac : Integral perturbed angular correlation.

Ipma : Ion probe microanalysis.

Iridium anomaly : The name given to the sharp maximum observed in iridium abundance versus depth profile of certain geologic formations. Discovered in 1979, the interpretation remains controversial.

Irma : See immunoradiometric assay.

Irms : Isotope ratio mass spectrometer.

Irradiate : To expose to some form of radiation.

Irradiation : Exposure to ionizing radiation.

Isobars, nuclear : Nuclides having the same mass number but different atomic numbers.

Isodiapheres : Nuclei having equal neutron excesses.

Isomer : One of two or more nuclides having the same mass number A and atomic number

Z, but existing for measurable times in different nuclear energy states. OR Nuclides with the same number of neutrons and protons in different states of excitation.

Isomeric state : An excited nuclear state having a mean life long enough to be observed.

Isomeric transition : A spontaneous transition between two isomeric states of a nucleus or between an isomeric state and the ground state of a nucleus. OR A relatively long-lived radioactive decay in which a nucleus goes from a higher to a lower energy state. The mass number and the atomic number are unchanged.

Isomers, nuclear : See isomer.

Isorad : In nuclear geology, contours of equal radioactivity.

Isospin : In nuclear structure, a quantum mechanical formalism in which the charge on a nucleon is treated like the spin of an electron, that is, half-integral with two possible orientations. The 0 and +1 charges become +(1/2) and -(1/2) isospins.

Isotones : Nuclides having the same neutron number but different atomic numbers.

Isotope : Isotopes of a given element have the same atomic number (same number of protons in their nuclei) but different mass numbers (different number of neutrons in their nuclei). ^{238}U and ^{235}U are isotopes of uranium.

Isotope dilution : Mixing of a given nuclide with one or more of its isotopes.

Isotope dilution analysis : A method of quantitative analysis based on the measurement of the isotopic abundance of a nuclide after dilution by mixing with one or more of its isotopes.

Isotope dilution analysis, direct : Isotope dilution analysis used for the determination of a non-radioactive element with the aid of one of its radionuclides .

Isotope dilution analysis, reversed : Isotope dilution analysis used for the determination of the isotopic carrier in a solution of a radionuclide with the aid of one of its stable isotopes.

Isotope dilution analysis, substoichiometric : A method of isotope dilution analysis . The final isotopic abundance is estimated from the amount of the nuclide present in a quantity of the relevant element separated from the sample. That quantity must be identical to the quantity isolated from a standard solution of the radioisotope added. The analysis requires that the separations be substoichiometric, adding identical but smaller than stoichiometric amounts of the reagent to the standard and diluted sample solutions.

Isotope effect : The difference in behavior between two substances of which only the mass numbers of one or more of the constituent atoms are different.

Isotope exchange : Any process which does not lead to usual physicochemical changes of the system, but changes the distribution of the isotopes of the given element between different chemical forms, different phases, or inside molecules. The exchange of places between isotopes of atoms in different valency states, in different molecules or ions, or in different sites in the molecule or ion.

Isotope exchange analysis : A method of quantitative analysis based on the isotope exchange between isotopes of the element to be determined and other isotopes of this element in different valency state or in different molecules.

Isotope farm : A closed biological growth chamber in which biochemically labeled compounds are produced by growth in an isotopically enriched nutrient environment.

Isotope geochronology : Determining the age of geological materials using radiochemical or other isotopic dating techniques.

Isotope geology : Investigation of geological phenomena by means of stable and radioactive isotopes of elements and of changes in their abundance.

Isotope hydrology : The study of the properties and distribution of water on the surface of the earth, in the soil and rocks, and in the atmosphere as measured using stable and radioactive isotopes and measurements of their abundances.

Isotope ratio mass spectrometry : A mass spectroscopic instrument with low resolution but excellent precision able to determine very small changes in the isotopic composition of gas samples.

Isotope, stable : A nuclide which is not radioactive.

Isotopes : Nuclides having the same atomic number but different mass numbers.

Isotopic abundance : See abundance, isotopic .

Isotopic carrier : See carrier, isotopic.

Isotopic modification : Preparing a compound that has a macroscopic composition such that the isotopic ratio of nuclides for at least one element deviates measurably from that occurring in nature.

Isotopic neutron source : Neutron sources generated by a spontaneously fissioning nuclide or by a nuclear reaction induced by radiation from one radioisotope in a mixture.

Isotopic nuclides : Isotopes.

Isotopic separation : Operation for the purpose of modifying an isotopic abundance in a mixture of isotopes.

Isotopic substitution : A process in which all molecules have a specified nuclide at a specified position.

Isotopic tracer : See tracer, isotopic.

Isotopomer : Chemically identical structures that differ in the atomic mass of one or more of their elements. J.

Isotron : A device for sorting isotopes of an element in which ions are accelerated to a fixed energy in a strong electric field and an radiofrequency field then selects ions according to their velocity which is inversely proportional to the square root of their mass.

It : Isomeric transition .

Ivnaa : In vivo neutron activation analysis.

Ivpgaa : In vivo prompt gamma neutron activation analysis.

Joule (J) : Unit of energy, equivalent to the work done in lifting a one-newton weight a distance of one meter.

Kbq : thousand Becquerels

K-capture : The capture by an atom's nucleus of an electron from the innermost electron orbital (K-shell) surrounding the nucleus.

K-electron capture : Radioactive decay of a nucleus by electron capture in which the electron captured by the nucleus was originally in the K-shell (1s atomic orbital).

Kelvin (K) : Unit of temperature equal in size to the Celsius degree, but with the zero set by the absolute zero of temperature, -273.15°C. Ice freezes at 273 K, room temperature is about 293 K, and water boils at 373 K, at sea level. human body temperature is 310K.

Kerma : Kinetic energy released in material.

Kev : kiloelectronvolt; 103 electronvolts.

Kev : Thousand electron volts.

Kev : One thousand electron-volts.

Km : kilometer(s).

Ko factor	: In activation analysis , the ratio of the product of atomic weight, isotopic abundance of activated nuclide, abundance of gamma-ray being measured, and the 2200 m/s cross section for an analyte element and a monitor nuclide. Its use in the ko method enables independence from irradiation conditions (thermal-to-epithermal ratio) and from counting conditions. The factor can be determined experimentally or from tabulated constants.
Kpa	: KiloPascals
Kryptonates	: A solid compound or metal into which ^{85}Kr has been incorporated by ion bombardment or thermal
Kt	: Kiloton(s)
Kvp	: Kilovolts (peak)
Label	: A marker, tag or indicator distinguishable by the observer but not by the system and used to identify a tracer.
Labeled	: Made identifiable by a label.
Labeling	: The act of incorporating a label into a chemical.
Labeling, (site) specific	: Isotopic labeling at a definite location on a complex molecule.
Labeling, autologous	: The radiolabeling of cells or plasma from a patient before re-administration to the same patient.
Labeling, cold	: Chemical or biochemical synthesis which incorporates a preformed label into a desired product.
Labeling, conjugation	: Labeling of a substance by conjugation with a labeled molecule.
Labeling, end	: Procedure in which terminal phosphate groups of DNA are removed and enzymatically replaced with radiolabeled groups.

Labeling, exchange : Labeling of a substance by isotope exchange.

Labeling, general : Labeling in which compounds are nonuniformly labeled where the position of the radioactive atom in the molecule varies from one molecule to the other. Also called random labeling.

Labeling, isotopic : Labeling in which the resulting product is only different from the initial one by its isotopic composition.

Labeling, non-isotopic : Labeling in which the resulting product has a different chemical composition from the initial one.

Labeling, recoil : Labeling by a chemical reaction initiated by recoil.

Labeling, wilzbach : Labeling of a substance by exposing it to tritium gas.

Lanthanides : An element which is any of the first series of f-block elements or inner transition elements commencing with lanthanum at atomic number 57 and ending with lutetium at atomic number 71.

Lanthanoids : See lanthanides.

Laser optoacoustic spectroscopy : An optical spectroscopy technique in which an optoacoustic modulator is used to produce an equidistant series of sharp pulses in a laser beam.

Latence time (of a detector) : See dead time .

L-capture : See L-electron capture.

LD : Lethal dose

Leach rate : The rate at which radioactivities are leached from solid waste aggregates by solubilization in water or aqueous liquid.

Lead castle, cave or housing : A structure, made of lead bricks, to enclose small radioactive sources. The structure may or may not contain viewing or handling facilities.

Lead equivalent : The thickness of lead, which, under specified conditions or irradiation, affords the same protection as the material under consideration.

Lead glass : Glass into which lead oxide is incorporated. Used for shielding purposes when limited visual observation is still necessary.

Leakage : In reactors, the loss of neutrons to the surroundings. In general, all radiation except the useful beam.

Least squares technique : A procedure for replacing the discrete set of results obtained from an experiment by a continuous function which best approximates the discrete results.

L-electron capture : Radioactive decay of a nucleus by electron capture in which the electron captured by the nucleus was originally in the L-shell (2s or 2p atomic orbital).

Lepd : Low-energy photon detector.

Lepton : A particle (such as the electron or neutrino) not subject to strong interactions.

Lepton number : Additive quantum number defining leptons; the three lepton numbers are electron number, muon number, and tau number. These numbers remain the same in all reactions.

Let : Linear energy transfer.

Lethargy (of a neutron) : A measure of the energy lost by neutrons in the slowing down due to multiple scatterings in the moderator.

Lewis effect : A peak in the thick target yield curves of narrow resonances near the resonance energy, due to the quantized character of charged-particle energy loss in penetrating matter.

Lifetime : The mean life of a particle or radioactive nucleus. This is equivalent to the decay time.

Ligand : A substance or part of a substance that binds to a specific receptor.

Linac : A linear accelerator .

Linac : Another name for a linear accelerator.

Linear accelerator : Particle accelerator laid out in a straight line. *OR* An accelerator in which the beam travels in a straight line during the entire acceleration process.

Linear analyzer : A position sensitive detector, used in radiochromatography imaging, in which there is no motion of detector relative to sample.

Linear electron accelerator : An evacuated metal tube in which electrons pass through a series of small gaps (usually in the form of cavity resonators in the high frequency range) so arranged and spaced that, at a specific excitation frequency, the stream of electrons on passing through successive gaps gains additional energy from the electric field in each gap.

Linear energy transfer : The average energy locally imparted to a medium by a charged particle of specified energy, per unit distance traversed.

Linear pulse amplifier : See amplifier, linear pulse .

Liquid drop model : A model of the nucleus in which the qualitative similarities to the cohesive forces of an ordinary liquid are recognized. The effects of volume energy and surface effects are supplemented by a variety of more subtle effects of both classical and quantal nature.

Liquid emulsion : Nuclear (photographic) emulsion manufactured in gel form so that after heating it can be poured to the desired shape at the desired time.

Liquid ion-exchanger : See ion-exchanger, e.g., di-(2-ethylhexyl) phosphoric acid (HDEHP) is a liquid cation exchanger.

Liquid scintillation counter : A radiation counter utilizing a liquid scintillation detector .

Liquid scintillation detector : See detector, liquid scintillation .

Lithium-drifted detector : See germanium detector, Ge(Li).

Live time : For a measurement, the time during which a radiation measuring assembly is capable of processing events occurring in the radiation detector . It equals the clock time minus the integrated resolving or dead time.

Lixiviation : Extraction from a solid mixture by washing or percolation.

Llod : Lower limit of detection; synonymous with limit of detection.

Lod : Limit of detection.

Logit : In competitive binding assays , the logit-log dose relationship, in which the response is defined by R=logit(y)=log[y/(1-y)] where y = b/bo with b = fraction of tracer bound and bo = value of b with no unlabeled ligand in the system. Logit transformed assay data frequently yield straight-line dose-response curves, amenable to statistical analysis.

Loss free counting : A scheme for correcting for dead time losses. It uses a pulser in conjunction with calculating the number of lost pulses during a dead period to augment the next spectral pulse artificially.

Low level counting : Any counting system in which special care is used to optimize the counting efficiency and reduce the background count rate. Such features as anticoincidence shielding and special materials are usually employed.

Lpo method : A radioiodination procedure employing the enzyme lactoperoxidase.

Luminescence : A phenomenon in which the absorption of primary radiation by a substance gives rise to the emission of electromagnetic radiation characteristic for that substance.

M : meter(s)

Mssbauer nuclide : Any radioactive nuclide that may serve as a radiation source in Mssbauer effect experiments.

Macrocycles : An organic molecule with a large ring structure, usually containing over fifteen atoms.

Macroporous ion exchanger : An ion exchanger with pores that are large compared to atomic dimensions.

Magic numbers : In the shell model of nuclear structure, nuclei with certain neutron numbers and certain proton numbers are observed to be particularly stable. The values of these so-called "magic" numbers are 2, 8,20, 28, 50, and 82. A higher shell at neutron number 126 exists and there are theoretical predictions about additional shells for protons and neutrons.

Manipulator : A hand-operated or -controlled device for remotely handling radioactive materials.

Man-made element : See artificial element .

Masking agent : A substance preventing the interfering reaction of one or more foreign substances in a determination by conversion into soluble complexes, different oxidation states, or other unreactive forms.

Mass defect : The difference between the sum of the rest masses of protons and neutrons which form a nucleus and the measured mass of the same nucleus.

Mass energy : Energy a particle has by virtue of its mass.

Mass excess : The excess of the measured mass of an atom over its mass number .

Mass number : The total number of neutrons and protons in a nucleus. Synonymous with nucleon number. ***OR*** The total number of protons and neutrons in the nucleus A=Z+N. This is also the total nucleon number of the nucleus.

Mass spectrometry : The use of instruments that analyze ions by mass-to-charge ratio to measure abundances of isotopes .

Mass-energy equivalence : The equivalence of mass m and energy E as expressed in Einstein's equation $E = mc^2$.

Matrix (in analysis) : In analysis: refers to the analytical sample, considered as an assemblage of constituents, with all their individual-properties. The combined effect that the various constituents of the matrixmay exert on the measure of the analysis of the element is referredto as the matrix effect.

Maximum permissible concentration : An expected upper limit for the concentration of a specified radionuclide in a material taken into the body below which continuous exposure (or in the case of occupational maximum permissible concentration, exposure for 40h/wk) to the material is not considered biologically harmful.

Mca : Multi channel analyzer. See pulse height analyzer, multi channel.

Mean free path : The average distance traveled by a particle between collisions.

Mean life, radioactive : The average lifetime of a radioactive nuclide.

Mean lifetime : See average life.

Mean linear range : See range, mean linear.

Mean mass range : See range, mean mass.

Measured spectrum : See spectrogram .

Mechanism : A detailed description of the pathway leading from the reactants to the products,

	including as complete a characterization as possible of the composition, structure and other properties of reaction intermediates and transition states.
Meson	: A particle (such as the pion) made of quark-antiquark pairs.
Metastable state	: An isomeric state in an energy state higher than the ground state.
Mev	: Milli electron volt = 1/1000 electron volt.
Mibk	: Methylisobutylketone, a common solvent used in extraction separations.
Microautoradiography	: Autoradiography for viewing with a microscope.
Micro-pixe	: Particle-induced x-ray emission spectroscopy in which the particle beam is highly collimated and focused allowing position sensitive analysis.
Microscopy, nuclear	: Use of nuclear microprobes to investigate the elemental structure of biological objects.
Microtron	: A cyclotron for accelerating electrons.
Microwaves	: Electromagnetic radiation with wavelength intermediate between radio wave and infrared radiation.
Milking	: The process of separating a radioactive tracer from a parent radionuclide bound in a radioisotope generator .
Mineralization	: 1. Formation of mineral in vivo, e.g., in bone, tooth, etc. 2. Use of reagents and various procedures to completely solubilize a solid sample mixture for analysis. (Not recommended.)
Mirror nuclei	: A pair of isobars in which the atomic and neutron number of one equals the neutron and atomic number of the other, respectively.

Mixed labeled : A specifically labeled compound is referred to as mixed labeled when the isotopically substituted compound has more than one kind of modified atom.

Mm : millimeter(s)

Mm^3 : cubic millimeters

Mnaa : Molecular neutron activation analysis.

Moderation : Reduction of the neutron energy by scattering without appreciable capture.

Moderator : A material used to reduce neutron energy by moderation).

Modification, isotopic : A process in which a compound has been given an abnormal isotopic ratio in one or more elements present.

Molecular activation analysis : See molecular neutron activation analysis .

Molecular neutron activation analysis : A variation of radiochemical neutron activation analysis in which chemical speciation (often by bioanalytical techniques) is involved in pre-irradiation separation. Synonymous with chemical neutron activation analysis, which is preferred.

Molecular weight : Also known as the relative molecular mass of a substance, the molecular weight is the ratio of the average mass per formula unit of a substance to 1/12 of the mass of an atom of nuclide 12C.

Monitor, beam : See beam, monitoring.

Monitor, flux : See flux monitor.

Monoenergetic radiation : Radiation consisting of particles of a single kinetic energy or photons of a single energy or wavelength.

Mother : In radiochemistry, synonymous with parent . See parent.

Mother liquor : The liquid that has passed through a chemical process.

Mpc : Maximum permissible concentration.

Mssbauer effect : Resonance absorption of gamma radiation by nuclei arranged in a crystal lattice in such a way that the recoil momentum is shared by many atoms.

Mssbauer spectrometry : The spectroscopic technique based on the Mssbauer effect .

Mt : megaton

Mu : The symbol for muonium used to emphasize the analogy of muonium to atomic hydrogen.

Multichannel (pulse height) analyzer : See pulse height analyzer, multi channel .

Multiple scattering : Successive rescattering of radiation within the scattering medium.

Multiplication neutron : The process in which a neutron produces on the average more than one neutron in a medium containing fissile material.

Multiscaling : A mode of operation of a multi channel pulse height analyzer for studying the time dependence of radiation. The analyzer, when operated in the "multiscaling" mode, stores all pulses in a fixed channel until a preset time, the "dwell time", elapses, after which the pulses are stored in the next channel, and so on. Dwell times are typically microseconds or larger.

Multiwire proportional counter : Particle detector using changes in the current in wires due to the passage of ionizing particles nearby.

MU-mesic atom or **molecule** : An atom or molecule which has captured a negatively charged muon into a stable atomic or molecular orbital.

Muon : Mu-meson; a fundamental particle related to the electron.

Muon : A charged lepton about 200 times more massive than an electron.

Muon number : Additive quantum number characterizing muons and muon neutrinos.

Muonium : An atomic system consisting of a muon and an oppositely charged electron .

N : 1. Neutron number . 2. In dosimetry, the factor designed to take into account all parameters influencing a dose equivalent , except the absorbed dose and type and energy of the radiation.

Na : Sodium

Naa : Neutron activation analysis .

Nai(Tl) detector : A scintillation crystal made of thallium-activated sodium iodide and used as a scintillation detector .

Nato : North Atlantic Treaty Organization

Natural isotopic abundance : See abundance, natural isotopic .

Natural radioactivity : Radioactivity of naturally occurring nuclides in materials where the isotopic abundance of that nuclide is natural.

Nca : No carrier added.

Ndp : (1). Neutron diagnostic probe. (2). Neutron depth profiling.

Negatron : A negative beta-particle .

Neutrino : Any of several neutral, possibly massless particles always involved with electrons in beta-decay nuclear transformations. *OR* An electrically neutral particle with negligible mass. It is produced in processes such as beta decay and reactions that involve the weak force. *OR* One of the basic particles that make up a nucleus. A neutron and a proton have about the same mass, but the neutron has no electrical charge.

Neutron activation : See activation.

Neutron activation mass spectrometry (nams) : A technique in which the products of activation are detected by mass spectrometry. For example, in determining boron in biological samples, neutron activation produces alphas by the ^{10}B reaction and the alphas are subsequently measured as ^{4}He in the mass spectrometer.

Neutron capture : A nuclear reaction in which a neutron is absorbed by a target nucleus producing an isotope one mass number greater in its ground or excited states.

Neutron density : The number of free neutrons divided by the containing volume. Partial densities may be defined for neutrons characterized by such parameters as energy and directions.

Neutron depth profiling : A method of charged particle activation analysis in which deuterons of specific energy undergo (d,n) reactions in which the neutron energy spectrum is measured by time-of-flight techniques. The energy is a measure of the deuteron's energy when the reaction occurred and therefore a probe of how deep the reacting element is in the sample. The intensity of detected neutrons is a measure of reacting element concentration.

Neutron diagnostic probe : A device for prompt analysis that is a combination of a neutron generator , an alpha-particle counter for time-of-flight neutron spectrometry, and a detector for inelastic gamma-ray spectroscopy.

Neutron excess : The number of neutrons in excess of the number of protons in a nuclide .

Neutron flux : See flux density, particle .

Neutron generator : 1.) An accelerator, usually small, of light charged ions (p, d, t) which produces a flux of neutrons through nuclear reactions in a light element target. 2.) Any device which produces neutrons.

Neutron number : The number of neutrons in a nucleus. OR The total number of neutrons in the nucleus, N.

Neutron temperature : The temperature assigned to a population of neutrons when this population is approximated by a Maxwellian distribution.

Neutrons, cold : Neutrons with a neutron temperature considerably lower than normal room temperature.

Neutrons, delayed : Neutrons emitted by fission products formed by nuclear decay (the observed delay is due to the lifetime of the preceding nuclear decay or decays.

Neutrons, epicadmium : Neutrons of kinetic energy greater than the effective cadmium cut-off for neutrons.

Neutrons, epithermal : Neutrons of kinetic energy greater than that of thermal agitation. The term is often restricted to energies just above thermal.

Neutrons, fast : Neutrons of kinetic energy greater than some specified value. This value may vary over a wide range and will be dependent upon the application, such as reactor physics, shielding or dosimetry.

Neutrons, fission : Neutrons originating in the fission process which have retained their original energy.

Neutrons, intermediate : Neutrons of kinetic energy between the energies of slow and fast neutrons . In reactor physics, the range might be 1 eV to 0.1 MeV.

Neutrons, prompt : Neutrons accompanying the fission process without measurable delay.

Neutrons, resonance : Neutrons , the energy of which corresponds to the resonance energy of a specified nuclide or element . If the nuclide is not specified, the term refers to resonance neutrons of ^{238}U.

Neutrons, slow : Neutrons of kinetic energy less than some specified value. This value may vary over a wide range and depends on the application. In reactor physics, the value is frequently chosen to be 1 eV; in dosimetry , the effective cadmium cut-off is used.

Neutrons, thermal : Neutrons in thermal equilibrium with the medium in which they exist.

Nick translation : Labeling procedure in which DNA is first "nicked" (broken) and a radioactive nucleotide is re-included by the action of DNA polymerase.

NIM module : An electronic component for nuclear instrumentation packaged in a standardized way to allow inter changeability with alternative units and to provide compatibility with equivalently standardized housing and powering units (bins) called NIM bins .

Nimbin : Standard bins for housing and supplying power to complementarily standardized electronics called nuclear instrument modules, or NIM modules .

No carrier added : A preparation of a radioactive isotope which is essentially free from stable isotopes of the element in question.

Non-activation analysis : Methods of analysis based on the interaction of nuclear radiation or X-rays with the sample, leading to absorption and backscattering, not to activation of the determined component.

Non-destructive activation analysis : See activation analysis.

Non-ionizing radiation : Radiation , the energy of which is insufficient to produce ions in matter. Included are microwaves, infrared light, visible light and slow neutrons. In most cases, ultraviolet radiation may be considered a non-ionizing radiation too.

Non-specific binding : That portion of the tracer used in a competitive binding assay that is found in the bound fraction, independent of the binding reaction. Also expressed as minimum binding.

Nra : Nuclear reaction analysis.

Nsb : See non-specific binding.

Nuclear activation analysis : See activation analysis.

Nuclear binding energy : The energy that free nucleons give up in order to be bound inside a nucleus.

Nuclear chemistry : See chemistry, nuclear .

Nuclear decay : See decay, nuclear .

Nuclear emulsion : A photographic emulsion specially prepared for the detection of nuclear particles. Such emulsions are generally thicker and contain a higher concentration of silver halide than those used for standard photography. The emulsion is usually deposited on a glass plate.

Nuclear fission : See fission, nuclear.

Nuclear force : See force, nuclear.

Nuclear fusion : See fusion, nuclear.

Nuclear geology : Application of nuclear science and techniques to geologic studies.

Nuclear isomers : See isomers, nuclear.

Nuclear level : One of the energy values at which a nucleus can exist for a measurable time (>10-22 s).

Nuclear medicine : A branch of medicine "taken to embrace all applications of radioactive materials in diagnosis or treatment or in medical research with the exception of the use of sealed radiation sources in radiotherapy". (World Health Organization, 1972); "the clinical and laboratory medical specialty that employs for diagnosis, therapy, and research the nuclear properties of radioactive and stable nuclides to evaluate metabolic, physiologic, and pathologic conditions of the body" (American Board of Nuclear Medicine); "devoted to the diagnosis, treatment, research, and prevention of disease by making use of unsealed radioactive sources and of the properties of stable radionuclides"

(Council of Europe and the Society of Nuclear Medicine Europe). Direct therapy using accelerators is also included.

Nuclear microprobe : A charged-particle or ion beam with excellent spatial resolution and used for nuclear reaction analysis.

Nuclear migration : Transport of nuclides from their point of origin by natural processes.

Nuclear particle : See particle, nuclear.

Nuclear pharmaceuticals : See radiopharmaceuticals.

Nuclear radius : The geometrical radius of the atomic nucleus; because of the relative incompressibility of nuclear matter, it is approximately proportional to the cube root of the atomic number of the nucleus.

Nuclear reaction : A transformation of a nucleus which occurs (i) when another nuclear particle approaches closely enough to be influenced by the mutual attractive nuclear force between them or (ii) when affected by an electromagnetic force of sufficient strength.

Nuclear reaction analysis : Charged-particle induced nuclear reactions for determination of elements, usually depth profiling of light elements.

Nuclear reactor : A device in which a fission chain reaction can be initiated, maintained, and controlled. Its essential components are fissionable fuel, moderator, shielding, control rods, and coolant.

Nuclear resonance broadening : A technique used for obtaining concentration depth profiles by comparing resonant charged-particle capture gamma-ray yields from an unknown distribution in a sample to those yields from a known distribution of the same isotope. Frequently, the latter distribution is just the natural isotopic mixture.

Nuclear resonance scattering : Mossbauer effect.

Nuclear solid state track detector : See solid state track detector.

Nuclear spectroscopy : The measurement of the energy spectrum of particles emitted or scattered from nuclei.

Nuclear spin : See spin, nuclear .

Nuclear transformation : See transformation, nuclear .

Nuclear transition : See transition, nuclear .

Nucleide : Nuclide is preferred.

Nucleon : A proton or a neutron. ***OR*** A constituent of the nucleus; that is, a proton or a neutron.

Nucleon number : Number of nucleons in a nucleus.

Nucleonics : The practical applications of nuclear science and the techniques associated with these applications.

Nucleopore filter : A filter with micron and submicron sized pores. The pores are of controlled and uniform size, produced by chemically etching radiation damage tracks of heavy ions with which the filters are bombarded as part of the production process.

Nucleus : The positively charged central portion of an atom , excluding the orbital electrons. ***OR*** The core of the atom, where most of its mass and all of its positive charge is concentrated. Except for ^{1}H, the nucleus consists of a combination of protons and neutrons.

Nuclide : A species of an atom characterized by its mass number, atomic number and nuclear energy state, provided that the mean life in that state is long enough to be observable. ***OR*** Any species of atom that exists for a measurable length of time. A nuclide can be distinguished by its atomic mass, atomic number, and energy state.

Nuclidic mass : The rest mass of a nuclide expressed in atomic mass units.

Off-line : Reference to a procedure or measurement done in conjunction with an accelerator or reactor but neither at the reaction site nor at the exact time the reaction occurs.

Oklo reactors : Naturally occurring nuclear reactors which operated at the Oklo mine in Gabon, Africa nearly 2 X 109 years ago. Geological uranium deposits partially depleted of fissile uranium. In the past, the ores were substantially more enriched in fissile uranium. The combination of the richness of the ore, its higher ^{235}U content (~3%) at the time, and presence of water (to act as moderator) caused the formation to behave as a natural fission reactor; burning fuel, producing fission radionuclides , and boiling off water, the latter resulting in "shut-down". This behavior can cycle.

On-line : Reference to a procedure or measurement done in conjunction with an accelerator or reactor and performed at the reaction site or at the exact time the reaction occurs.

P.I.N semiconductor detector : See semiconductor detector, P.I.N.

Paa : Photon activation analysis. See instrumental photon activation analysis .

Pac : Perturbed angular correlation .

Packing fraction : The mass excess of a nucleus divided by the mass number; that is, the average mass excess per nucleon.

Paes : Positron-annihilation-induced Auger-electron emission spectroscopy.

Pair attenuation coefficient : The attenuation coefficient when only the pair production process is taken into account.

Pair production : The simultaneous formation of an electron and a positron as a result of the interaction of a photon of sufficient energy (> 1.02 MeV) with the field of a charged particle .

Paralysis time : See dead time .

Parent : A radionuclide that decays to another nuclide. *OR* A nuclide which is a precursor to another nuclide in a decay chain.

Parity : A symmetry property of a spatial wave function. The parity is said to be even (or +) if the wave function is left unchanged by reversing the sign of all the coordinates, odd (or -) if the sign of the wave function is thereby changed.

Partial cross section : The portion of the total cross section corresponding to a particular final reaction channel, that is, to a distinct set of reaction products.

Partial decay constant : For a radionuclide , the probability in unit time for the decay of one of its nuclei by one of several modes of decay.

Particle : A nuclear or an elementary particle.

Particle density : The number of particles divided by the containing volume.

Particle flux density : See flux density, particle .

Particle, elementary : A particle in which, at the present, no structure can be observed at moderate energies.

Particle, nuclear : A nucleus or any of its constituents in any of their energy states.

Particle-induced desorption mass spectrometry : Time-of-flight mass analysis of a sample that is desorbed from a surface as a result of bombardment by micro beams of charged particles , most often heavy-ions, for instance with a nuclide decaying by spontaneous fission .

Particle-induced gamma emission : Charged particle activation analysis in which the gamma radiations are detected on line) from the electromagnetic de-excitation of the product nuclei.

Particle-induced prompt photon spectrometry : See particle-induced gamma emission .

Particle-induced x-ray emission spectroscopy : Any technique in which a specimen is bombarded with a focused beam of high-energy particles and the characteristic x-ray spectrum generated in the specimen is recorded.

Partition coefficient : Distribution ratio of a radionuclide between two phases.

Passive assay or **interrogation** : A non-destructive measurement of nuclear material content or element or isotopic concentration of an item carried out by observing spontaneous radiations .

Pbq : Peta-becquerel (10^{18}Bq)

Pd : Performance decrement

Pdms : Particle-induced desorption mass spectrometry.

Peak analysis : The extraction of relevant peak parameters (i.e., position, area) from a measured spectrum .

Peak area method : A method of peak analysis in which a peak area is calculated by subtracting an estimate of the underlying continuum in a relevant part of a measured spectrum.

Peak fitting : A method of peak analysis in which a relevant part of a spectrum is fitted with a theoretical response function.

Peak-to-compton ratio : In a gamma-ray spectrum, the relative heights of the full energy peak to the broad continuum that lies below its Compton edge . It is a measure of a detector's ability to distinguish full energy peaks from the general smooth background in a complex spectrum.

Pellicle : In nuclear emulsions , the thin layer or film of photographic emulsion material.

Penetration factor : The fraction of particles (or probability per particle) which pass through a classically forbidden barrier. The barrier can be due to centripetal and Coulomb effects, for example.

Perturbed angular correlations : Angular correlations between coincident radiations in which the angular correlation coefficients have a time-dependent factor which depends on the lifetime of the intermediate state and on the interaction of the nuclear electromagnetic moments with the local atomic or molecular electric field.

Pesa : See proton elastic scattering analysis .

Pet : Positron emission tomography

Pgaa : Prompt gamma-ray activation analysis. See prompt gamma-ray neutron activation analysis .

Pgnaa : Prompt gamma-ray neutron activation analysis .

Pha : Pulse height analysis or analyzer .

Photoactivation : Activation induced by photons, that is, by electromagnetic radiation.

Photoeffect : See photoelectric effect .

Photoelectric attenuation coefficient : The attenuation coefficient when only the photoelectric process is taken into account.

Photoelectric effect : The complete absorption of a photon by an atom with the emission of an orbital electro . Synonymous with photoeffect.

Photoelectric peak : Of a spectrum of gamma radiation , the part of the spectral response curve corresponding to the absorption in the radiation detector by the photoelectric effect of the detected gamma energy. In most cases the peak also contains the events caused by multiple processes and use of the expressions total absorption peak or full energy peak is to be preferred. Synonymous with photopeak.

Photoelectron : The electron ejected from an atom which completely absorbs a photon in the photoelectric effect .

Photofraction : The fraction of detected gamma-rays that produce output signals in the photopeak region.

Photomultiplier : Commonly used device for detecting photons by converting them to an electrical signal.

Photomultiplier tube : A vacuum tube containing a photosensitive layer which serves as the cathode for an electron multiplier.

Photon : A quantum of electromagnetic radiatio. A packet of electromagnetic energy. Photons have momentum and energy, but no rest mass or electrical charge.

Photon activation : See activation.

Photoneutron : A neutron which is ejected from a nucleus as a consequence of the absorption of a photon .

Photonuclear reaction : A nuclear reaction induced by absorption of a photon .

Photopeak : See photoelectric peak .

Photopeak efficiency, absolute : Of a gamma-ray spectrometer, the counting efficiency when only considering the events recorded in the photopeak.

Photopeak efficiency, intrinsic : The detector efficiency when only considering counts resulting from the photoelectric effect.

Pig : A small shielding container for storing or carrying radioactive material or for housing a detector system for measuring radioactivity.

Pige : Particle-induced gamma emission.

Pigme : See particle induced gamma emission.

Pile : Obsolete term for a nuclear reactor.

Pile-up : The processing by a radiation spectrometer of pulses resulting from the simultaneous absorption of independent particles or photons in a radiation detector . As a result they are counted as one single particle or photon with energy between the individual energies and the sum of these energies.

Pion : The least massive known spin-0 meson. The three charge states of the pion (negative, neutral and positive) are involved in the long-range force between the nucleons.

Pipps : Particle-induced prompt photon spectroscopy.

Pixe : Particle-induced x-ray emission .

Plasma desorption mass spectrometry : See Particle-induced desorption mass spectrometry.

Plateau : The plot of counting rate against voltage applied to a Geiger counter in the presence of a constant source of radiation. As the voltage is increased from zero, this counting rate first rises sharply to a plateau several hundred volts long. Over the plateau the counting rate rises very slowly, and at the end it increases rapidly as instability sets in.

Pm : Photomultiplier tube.

Pneumatic tube : A cylindrical pipe for transporting a rabbit (sample) by means of compressed air.

Pocket ion chamber : A small dosimeter having a physical appearance similar to a fountain pen, used to monitor the amount of radiation to which an individual is exposed.

Popop : diphenyloxazolbenzene, used in liquid scintillation spectroscopy as a wavelength shifter.

Position sensitive detector : A radiation detector which records not only physical properties of the radiation, but also where it impinges on the detector surface.

Positron : A positively charged electron.

Positron annihilation : The conversion of a positron and its antiparticle, the electron, into annihilation radiation.

Positron decay : Radioactive decay in which the emission of a positron by the decaying nucleus occurs, accompanied by an emitted neutrino.

Positron emission tomography : An external technique for the rapid serial reconstruction of the spatial distribution of any positron-emitting radioisotope that has been administered in vivo .

Positron-annihilation-induced auger-electron emission spectroscopy : A form of Auger electron emission spectroscopy based on the use of thermalized positrons whose annihilation excites the Auger emission. The surface analysis technique is particularly sensitive to the outermost few angstroms layer.

Positronium : The short-lived atom consisting of a positron and electron.

Postlabeling : A method for determination and measurement of covalent DNA adducts based on enzymatic incorporation of ^{32}P into DNA nucleotides.

Power, stopping : See stopping power .

Ppo : 2,5-diphenyloxalol, used in liquid scintillation spectrometry as a primary scintillator.

Pra : Prompt radiation analysis. Use prompt gamma neutron activation analysis .

Preamplifier : An amplifier whose primary function is boosting the output of a low-level frequency source to an intermediate level so that the signal may be further processed without appreciable degradation of the signal-to-noise ratio of the system.

Precision : A measure for the reproducibility of measurements within a set, that is, of the scatter or dispersion of a set about its central value.

Preconcentration : In trace analysis: a process resulting in an increase in the concentration or the amount of the components of interest. As a synonym, enrichment is not recommended.

Precursor : Of a nuclide , any radioactive nuclide which precedes that nuclide in a decay chain.

Projectile : A particle in a beam being used for irradiation .

Prompt gamma : A gamma ray which is emitted from an excited state of a nucleus immediately upon formation of that excited state by a previous decay step or reaction.

Prompt gamma neutron activation analysis : Neutron activation analysis in which the radiation of the neutron capture gamma-ray cascade is detected on-line , that is, simultaneously with the production.

Prompt neutron activation analysis : See prompt gamma neutron activation analysis .

Prompt radiation analysis : See prompt gamma neutron activation analysis .

Proportional counter (tube) : See counter tube, proportional .

Proton : One of the basic particles that makes up an atom. The proton is found in the nucleus and has a positive electrical charge equal to the negative charge of an electron and a mass similar to that of a neutron a hydrogen nucleus. *OR* A single nucleon, of mass number one, which is the nucleus of the hydrogen atom.

Proton elastic scattering analysis : Elastic recoil detection analysis using a proton beam.

Proton microbeam : A highly collimated and focused beam of protons.

Proton number : See atomic number .

Proton number : The total number of protons in the nucleus, Z.

Ps : The symbol for positronium used to emphasize the analogy of positronium to atomic hydrogen.

Pseudocyclic activation analysis : A variation of cyclic activation analysis in which a delay is introduced between irradiation and spectrum acquisition.

Pseudo-radiocolloid : An inactive colloid which incorporates radionuclides.

Pu : plutonium

Pulse amplitude analyzer : A sub-assembly for determining the distribution function of a set of pulses in terms of their amplitudes. Synonymous with pulse height analyzer.

Pulse amplitude selector : A circuit which gives an output pulse for each input pulse whose amplitude lies within a chosen interval. Synonymous with single channel pulse height analyzer.

Pulse height analysis : The process of determining the spectrum of radiation energies through use of a pulse height analyzer .

Pulse height analyzer : See pulse amplitude analyzer.

Pulse height analyzer, multichannel : A pulse amplitude analyzer which includes a storage function to record the number of pulses received per channel.

Pulse height analyzer, single channel : See pulse amplitude selector .

Pulsed neutrons : Neutrons of transient high flux density produced in certain reactors when control rods are removed.

Pulsed reactor activation analysis : Activation analysis using pulsed neutrons .

Purex process : A separation scheme for recovering plutonium and uranium from fission products in reactor fuel.

Purity, isotopic : See abundance, isotopic .

Purity, radiochemical : For a material, the fraction of the stated isotope present in the stated chemical form.

Purity, radionuclidic : Of a material, that fraction of the total activity which is present in the form of the stated radionuclide (including its daughter products).

Pwr : Pressurized water reactor.

Qcd : Quantum chromo dynamics, the gauge theory describing the color strong interaction.

Qed : Quantum electrodynamics, the gauge theory describing electromagnetism.

Qf : Quality factor .

Q_o-VALUE : The ratio of resonance integral to thermal cross section in activation analysis .

Quality factor : The linear-energy-transfer -dependent factor by which absorbed dose is multiplied to obtain dose equivalent.

Quality of radiation : Penetration of radiation: frequently measured by its half-value layer, the thickness of some standard material which transmits 50% of the incident radiation.

Quantum yield : The number of moles transformed in a specified process, physically (e.g., by emission of photons) or chemically, per mole of photons absorbed by the system.

Quark : A strongly-interacting fermion that is a building block of hadronic matter. Quarks come in six flavors up, down, charm, strange, top, bottom.

Quenching : 1. The process of inhibiting continuous or multiple discharges following a single ionizing event in certain types of radiation detectors), particularly in (Geiger-Muller counter tubes . 2. The deactivation of an electronically excited state by non-radiative process, e.g., as a source of error when using liquid scintillation detectors . 3. The shortening of the lifetime of positronium by its interaction with other species.

Quenching correction : Correction for errors due to different quenching for standards and test portions. When using liquid scintillation detectors , these corrections can be based on the standard addition or sample channels ratio method, or the use of automated external standardization.

Q-value : In a nuclear transformation, the energy equivalent of the difference in atomic masses between the reactants and products, all in their nuclear ground states.

R : Symbol for the roentgen .

Rabbit : A small container propelled pneumatically or hydraulically through a tube leading between the laboratory and a location in a nuclear reactor or other device where irradiation of a sample can take place.

Rad : A unit of absorbed dose . (Not a SI unit.) 1 rad is 0.01 J absorbed per kg of any material. 1 rad = .01 gray).

Rad : Radiation absorbed dose

Rad (radiation absorbed dose) : A former unit of an absorbed dose of ionizing radiation. One rad is equal to the absorption of radiation energy per gram of matter. It has been replaced by the gray (see above).

Radiation : A term embracing electromagnetic waves as well as fast moving particles.

Radiation chemistry : That part of chemistry which deals with the chemical effects of ionizing radiation , as distinguished from photochemistry associated with visible and ultraviolet electromagnetic radiation.

Radiation counter : See counter, radiation .

Radiation detector : See detector, radiation .

Radiation effects : Any of several effects brought about by the interactions of radiation with matter; effects such as radiolysis, ionization, excitation, and defect formation.

Radiation hazard : Hazard that exists in a region where there is a radiation field, other than what is considered to be normal background radiation.

Radiation inactivation method : A method for estimating protein size based on the correlation between the radiation dose applied and the amount of protein activity surviving the dose.

Radiation length : The distance traveled by a charged particle through matter over which its energy is decreased by a factor e.

Radiation source : An apparatus or material emitting or capable of emitting ionizing radiation.

Radiation spectrum : The components of radiation arranged in order of their wavelengths, frequencies or quantum energies. For particle radiation they are arranged in order of their kinetic energies.

Radiation, annihilation : See annihilation radiation .

Radiation, background : See background radiation .

Radiation, cerenkov : See Cerenkov radiation.

Radiation, hard : See hard radiation.

Radiation, ionizing : See ionizing radiation.

Radiation, monoenergetic : See monoenergetic radiation.

Radiation, natural : Radiation originating from natural radioactivity.

Radiation, penetrating : Gamma-rays, x-rays or neutrons with low interaction cross sections.

Radiation, soft : See soft radiation.

Radiative capture : See capture, radiative.

Radical : A molecular entity possessing an unpaired electron such as $\bullet CH_3$, $\bullet SnH_3$, $\bullet Cl$ (or $Cl\bullet$). (In these formulae the dot, symbolizing the unpaired electron, should be placed so as to indicate the atom of highest spin density, if this is possible.) Paramagnetic metal ions are not normally regarded as radicals.

Radioactive : The property of unstable nuclides of undergoing spontaneous nuclear transformations with the emission of radiation .

Radioactive age : Of an object, the time, estimated from measurement of the isotopic composition, during which the content of a radioactive species within that object has remained unchanged except for nuclear decay.

Radioactive chain : See decay chain.

Radioactive contamination : A radioactive substance in a material or place where it is undesirable.

Radioactive cooling : See cooling, radioactive .

Radioactive dating : A technique for estimating the age of an object by measuring the amounts of various radioisotopes in it.

Radioactive decay : See decay, radioactive .

Radioactive equilibrium : See equilibrium, radioactive.

Radioactive fallout : See fallout, radioactive .

Radioactive half-life : See half-life, radioactive .

Radioactive indicator : See indicator, radioactive .

Radioactive period : See radioactive mean life .

Radioactive series : See decay chain .

Radioactive source : Any quantity of radioactive material which is intended for use as a source of ionizing radiation.

Radioactive tracer : A tracer containing a radioactive label.

Radioactive tracer technique : In analysis: a technique for investigating recovery, loss, behavior, and speciation of a micro component, in which a radioactively-labeled element or compound chemically identical with the micro component is added to the sample, and its behavior is followed by radioactivity measurements.

Radioactive waste : Unwanted radioactive materials obtained in the processing or handling of radioactive materials. *OR* Materials that are radioactive and for which there is no further use.

Radioactivity : The property of certain nuclides of showing radioactive decay. *OR* The spontaneous decay or disintegration of an unstable atomic nucleus accompanied by the emission of radiation.

Radioactivity, artificial : See induced radioactivity.

Radioactivity, induced : See induced radioactivity.

Radioactivity, natural : See natural radioactivity.

Radioallergo(ad)sorbent test (rast) : An isotopic technique for the demonstration of reagins (a class of antibody) directed against specific allergens.

Radioanalytic imaging : A technique of separation and analysis of reaction products from a radiolabeled starting material. The products are separated by thin layer chromatography using an autoradiograph to quantitatively and qualitatively visualize the results. The technique can be used to study photo degradation, for example.

Radioanalytical chemistry : That part of analytical chemistry in which the application of radioactivity is an essential step in the analytical procedures. Synonymous with analytical radio-chemistry. See analytical radiochemistry.

Radioanalytical purification : See radiochemical purification.

Radioautograph : See autoradiograph .

Radiochemical purification : Chemical separation applied to a radioactive preparation in order to improve the radiochemical purity.

Radiochemical purity : See purity, radiochemical .

Radiochemical separation : Separation by a chemical means of the radioactive isotopes of (a) specified element(s) from a mixture of isotopes.

Radiochemical yield : 1) For the isotopes of a specified element, the yield of a radiochemical separation , expressed as a fraction of the activity originally present. 2) In radiation chemistry, the number of species transformed by radiation per 100 eV of absorbed energy: represented by the symbol G, the G-value .

Radiochemistry : That part of chemistry which deals with radioactive materials. It includes the production of radionuclides and their compounds by processing irradiated or naturally occurring radioactive materials, the application of chemical techniques to nuclear studies, and the application of radioactivity to the investigation of chemical, biochemical or biomedical problems.

Radiochemistry, analytical : That part of analytical chemistry in which the application of radioactivity is an essential step in the analytical procedures. Synonymous with radioanalytical chemistry.

Radiochromatogram : A graphical or other representation of a radiation detector response as a function of time produced by chromatographic separation of a mixture of radioactive species.

Radiochromatograph : A measuring assembly designed to measure the spatial or time distribution of the activity of a mixture of radioactive components after separation by a chromatographic method.

Radiocolloid : A radioactive substance in colloidal form. The radioactivity usually has little effect on the physicochemical properties of the colloid. See also pseudo radio colloid .

Radioecology : The scientific discipline dealing with the migration, transfer, and concentration of radionuclides in the environment and with the effect of environmental radioactivity on living organisms.

Radioelement : An element having no stable isotope. Also used to indicate an element which is naturally radioactive.

Radioenzymatic assay : An assay of the enzymatic activity of a substance based on the use of a radioactive substance

Radioexchange : An exchange reaction in which one of the exchanging species is radioactive .

Radiogenic : Resulting from radioactive decay, usually in reference to natural products such as lead.

Radiograph : A visual representation of an object produced by placing the object between a source of ionizing radiation and a photographic plate or film.

Radiogravimetric analysis : An analytical procedure in which the activity of a precipitate is used as a measure of its mass.

Radioimmunoassay : An analytical procedure based on the reversible and non-covalent binding of an antigen (hapten) by a specific antibody employing radioactively labeled antigen (hapten) to measure the fraction of the antigen (hapten) bound to a substoichiometric amount of antibody.

Radioimmunoassay, solid phase antibody : Method of radioimmunoassay employing an antibody bound to a solid phase.

Radioimmunoelectrophoresis : Immunoelectrophoresis in which either the antibodies or antigens are radioactively labeled.

Radioiodination : The process of incorporating the radionuclides of iodine (usually 125I, 131I or 123I) in, or of covalently linking a radioiodinated substance to, a substance.

Radioisotope : A radioactive isotope of a specified element. *OR* A radioactive isotope. A common term for a radionuclide.

Radioisotope dilution analysis : A method of isotope dilution analysis making use of a radionuclide.

Radioisotope generator : A system in which a long-lived radioisotope is permanently bound and decays to a shorter-lived daughter radioisotope that may be chemically removed or eluted. The latter process is referred to as milking the generator. The radioisotope generator is colloquially termed a "cow".

Radioligand : Radioactive ligand

Radioluminescence : Light emissions caused by radiations from ?radioactive materials.

Radiolysis : The chemical decomposition of materials by ionizing radiation.

Radiometric analysis : A method of analysis in which measurement of the activity is an essential step.

Radiometric assay : A non- competitive binding assay in which the receptor is labeled instead of the ligand. Synonymous with immunoradiometric assay .

Radiometric titration : A titration in which a radioactive indicator is used to monitor the end-point of the titration.

Radionuclide : A nuclide that is radioactive. OR A radioactive nuclide. An unstable isotope of an element that decays or disintegrates spontaneously, emitting radiation.

Radionuclidic purity : See purity, radionuclidic .

Radiopharmaceutical : A radioisotopically labeled pharmaceutical.

Radiopharmacology : The study of the preparation and properties of radioactive drugs and their therapeutic and diagnostic uses.

Radiopolarography : A tool for studying electrochemical phenomena in which labeled ions are reduced polarographically and the resulting drop is counted. The radioactivity of a given number of drops is proportional to the amount of the element deposited and also to the current. It is a low concentration technique.

Radioreagent : Reagent labeled with radioactivity .

Radioreceptor analysis : See radioreceptor assay .

Radioreceptor assay : Competitive binding assay employing a receptor protein that is not an antibody .

Radiorelease analysis : An analytical procedure based on the release of radioactivity from the reagent by reaction with the analyte .

Radiothermoluminescence : The reappearance of luminescence in? radioluminescent materials brought about by heating. Quartz and certain types of glass show the effect.

Radiotoxicity : The ability of a substance to give rise to adverse biological toxic effects as a result of the ionizing radiation it emits.

Radiotracer technique : See radioactive tracer technique .

Radwaste : Radioactive waste .

Raffinate : In solvent extraction between an organic phase and an aqueous phase, the aqueous phase after equilibrium has been established.

Random coincidence : See coincidence , random.

Range : The distance from a radiation source to the point where the flux density has vanished. A more explicit term is the mean linear range .

Range straggling : The stochastic dispersion of displacement (range) of charged particles about their mean linear range.

Range, extrapolated : See extrapolated range.

Range, mean linear : In a given material, for specified charged particles of a specified energy, the average displacement of the particles before they stop.

Range, mean mass : The mean linear range multiplied by the mass density of the material.

Range, residual : The penetration depth of a charged particle yet remaining to be traveled before the particle is stopped in the material.

Range-energy relation : The dependence of the mean linear range of a charged particle in a particular substance on the particle's initial energy.

Rare earth : See lanthanide .

Rast : See radioallergosorbent test .

Ratemeter : An electronic sub-assembly which gives a continuous indication proportional to the average counting rate over a predetermined time interval (time constant).

Rbe : Relative biological effectiveness.

Rbs : Rutherford backscattering spectroscopy.

Rcp : Radiochemical purity .

Reaction chain activation : See secondary particle activation analysis.

Reaction energy : See Q-value.

Reaction, nuclear : A transmutation or nuclear excitation induced by radiation from an external source. OR A device in which a self-sustaining nuclear fission chain reaction can be maintained and controlled. The term is sometimes applied to a device in which a nuclear fusion reaction can be produced and controlled.

Receptor : A chemical compound on a cell membrane or in a cell that binds other chemicals.

Recoil : The motion acquired by a particle through a collision with or the emission of another particle or electromagnetic radiation.

Recoil atom : The atom struck by a particle with which it is has undergone either elastic or inelastic scattering or reaction, or an atom or ion remaining after a radiation has been emitted.

Recovery : In trace analysis: the yield of a preconcentration step expressed as quantities of the concentrated component.. See also yield, radioactive.

Ree : Rare earth element(s); that is, lanthanides .

Reference material : A homogeneous material, available in sufficient amounts, which can be used for comparing analysis results obtained at different laboratories and/or with different techniques.

Relative atomic mass : See atomic mass, relative .

Relative biological effectiveness : The ratio of the appropriate value of the biological effectiveness of the radiation in question to that of x-radiation with an average specific ionization of 100 ion pairs per micron of water, for the particular biological effect under consideration and for the condition under which the radiation is received.

Relative counting : A measurement in which the activity of a sample is derived from the ratio between the count rates observed for the sample and for a source of known activity.

Rem : Roentgen equivalent, man OR A measure of dose deposited in body tissue, averaged over the body. One rem is approximately the dose from any radiation corresponding to exposure to one roentgen of radiation. The rem is no longer accepted for use with the International System. One rem is equivalent to 0.01 sievert.

Residual strong force : Force between composite objects (made of quarks) due to the remaining effect of the color force on colorless objects. These forces are much weaker than the strong color force.

Resolution : The use of the word resolution should be restricted to peak resolution . The concept should be distinguished clearly from that of resolving power, which should be restricted to instruments. There solving power of an instrument used in an analytical procedure may well have its impact on the selectivity of the overall procedure.

Resolution enhancer : 1) Detector designs that reduce degradation of resolution by compensating for ballistic effects) or charge trapping, 2) Mathematical procedures, usually called "filters", for improving the ability to accurately resolve close peaks.

Resolving power : The central position of the response curve of a spectrometer of monoenergetic radiation divided by the width at one half of its height.

Resolving time : The smallest time interval which must elapse between the occurrence of two consecutive ionizing events or signal pulses, in order that the measuring device be capable of detecting each of them separately.

Resolving time correction : Correction to be applied to the observed number of pulses in order to take into account the number of pulses lost during the resolving time .

Resonance : The enhancement of the response of a system to a periodic driving force when the driving frequency is equal to a natural frequency of the system. For nuclear systems, this manifests itself as a particular reaction having a cross section maximum.

Resonance energy : The energy of a particle entering a nuclear reaction, this energy being just sufficiently high to lead to the formation of reaction products in one of their excited states .

Resonance integral : The integral, over all or some specified portion of the resonance energy range, of the cross section divided by the energy of a radiation.

Rest mass : The mass of a particle in its ground state and at rest. It represents the Newtonian mass and does not include the additional mass acquired by a particle in motion due to the relativistic effect.

Retention : Of atoms undergoing a nuclear transformation, that fraction which remains in or reverts to its initial chemical form.

Reversed phase chromatography : A term of historical interest in liquid-liquid chromatography referring to an elution procedure in which the stationary phase is non-polar, e.g. paper treated with hydrocarbons or silicones.

Ria : Radioimmunoassay

Rnaa : Radiochemical neutron activation analysis. See activation analysis, radiochemical .

Roentgen : A unit of exposure. 1 roentgen = 1 R = 2.58 X 10^{-4} C/kg. (Not a SI unit).

Roentgen equivalent man (rem) : A unit of dose equivalent which depends on linear energy transfer and a quality factor . (Not a SI unit.) 1 rem = .01 J/kg = 1 rad X quality factor . 1 rem = 0.01 sievert (Sv).

Rpl : Radiophotoluminescent

Rra : Radioreceptor analysis .

Rsmr : Rayleigh scattering of Mossbauer resonance.

Rutherford : An obsolete unit of activity equal to 106 disintegrations per second.

Rutherford scattering or rutherford backscattering spectroscopy : Any technique using high-energy particles directed toward a sample, in which the bombarding particles, after scattering, are detected and recorded as a function of energy and/or angle. The technique is mostly used for determining the depth distributions of elements based on the energy of the backscattered particle.

Safeguards : Measures employed to prevent or detect the diversion of nuclear material and to protect against sabotage.

Salting out : Improving the extraction of a substance by the addition of an electrolyte to the aqueous phase.

Sargent curves or **sargent diagram** : Graphs of logarithms of radioactive disintegration constants of beta-emitting radioisotopes against corresponding logarithms of their maximum beta-particle energies.

Saturation : Of an irradiated element for a specified isotope , the steady state reached when the disintegration rate of the nuclide formed is equal to its production rate.

Saturation activity : For a specified isotope, the value of the activity of an irradiated element, when a state of saturation is reached.

Saturation analysis : A type of competitive binding assay where the specific reactor substance binding sites are all occupied (saturated) with ligand.

Saturation correction : Also known as saturation factor, the ?factor applied to an induced activity that transforms the activity into a hypothetical activity in an irradiation of infinite duration, that is, into a saturation activity.

Scaler : A sub-assembly for counting electrical pulses and containing one or more scaling circuits. ***OR*** An electronic instrument for counting radiation induced pulses from radiation detectors such as a Geiger-Muller tube.

Scaling circuit : An electronic circuit which produces an output pulse after a specified number of pulses has been received at its input.

Scatchard plot : In radioimmunoassays, for example, a graph of bound-to-free ratio vs the concentration of bound antigen.

Scattering : A process in which a change in direction or energy of an incident radiation is caused by interaction with a particle , a system of particles, or a photon.

Scattering recoil coincidence spectrometry : In depth profiling, a technique combining Rutherford backscattering spectroscopy and elastic recoil detection analysis . The recoil is detected in coincidence with the scattered beam particle.

Scattering, compton : See Compton effect.

Scattering, elastic : Scattering in which the total kinetic energy is unchanged.

Scattering, inelastic : Scattering in which the total kinetic energy changes.

Scavenger : A reagent used for scavenging.

Scavenging : In radiation chemistry : binding radicals or free electrons with a scavenger . In radiochemistry : the use of a precipitate to remove from solution by adsorption or coprecipitation, a large fraction of one or more radionuclides.

Scintigram : An image of the distribution of activity obtained with a scintillation camera following the internal administration of a radionuclide.

Scintillant : Scintillating material, fluor.

Scintillating material : Any substance constituting an appropriate medium for the detection of radiation by means of the scintillation phenomenon.

Scintillation : Burst of luminescence of short duration caused by an individual high-energy particle.

Scintillation cocktail : The solution of fluors used for liquid scintillation counting .

Scintillation counter : Radiation counter incorporating a scintillation detector .

Scintillation counter : An instrument that detects and measures gamma radiation by counting the light flashes (scintillations) induced by the radiation.

Scintillation crystal : A scintillation material that is crystalline in nature (as opposed to polymeric, for example).

Scintillation detector : A radiation detector using a medium in which a burst of luminescence radiation is produced along the path of an ionizing particle .

Scintillation spectrometer : A measuring assembly incorporating a scintillation detector and a pulse amplitude analyzer , used for determining the energy spectrum of certain types of radiation .

Scintillator : A finite quantity of scintillating material intended to be the element sensitive to radiation in a scintillation detector. ***OR*** Material that emits light when particles traverse it.

Scintiplast : Plastic scintillator.

Scission : The stage in the fission mechanism when the highly deformed nucleus divides into at least two fragments.

Scrubbing : In liquid-liquid distribution, the process of removing impurities from the separated phase containing the main substance.

Sd : Skin dose

Sealed source : A radioactive source sealed in a container or having a bonded cover, where the container or cover has sufficient mechanical strength to prevent contact with and dispersion of the radioactive material under the conditions of use and wear for which it was designed.

Sec : Second(s)

Secondary particle activation analysis : An activation method in which energetic particles from a primary reaction on a nuclide of interest subsequently undergo a secondary activation reaction on another nuclide producing a more easily detected radioactivity.

Secondary radiation : Radiation emitted by any matter irradiated with electromagnetic or ionizing radiation .

Secular equilibrium : A state of parent-daughter equilibrium that is achieved when the half-life of the parent is much longer than the half-life of the daughter. In this case, if the two are not separated, the daughter will eventually decay at the same rate at which it is being produced. At this point, both parent and daughter will decay at the same rate until the parent is essentially exhausted.

Segre curve : See beta stability line .

Self-absorption : The absorption of radiation by the emitter.

Self-absorption factor : Of a radiation source, the ratio between the amount of radiation emitted by the source and the amount of radiation produced by the radioactive nuclei present in the source. Synonymous with source efficiency .

Self-radiolysis : The radiolysis of a substance or mixture brought about by one of the substance's or mixture's own radioactive components.

Self-shielding : The lowering of the flux density in the inner part of a sample due to absorption in the outer layers of the sample.

Semiconductor : Material whose conductivity, due to charges of both signs, is normally in the range between that of metals and insulators and in which the electric charge carrier density can be changed by external means.

Semiconductor detector : Radiation detector using a semiconductor in which free electric charges are produced along the path of an ionizing particle.

Semiconductor detector, diffused junction : A semiconductor detector in which the P-N or N-P junction is produced by diffusion of donor or acceptor impurities.

Semiconductor detector, p.i.n. : A semiconductor detector consisting of a compensated region between a P and an

N region. The compensated region is often referred to as "intrinsic".

Semiconductor detector, surface barrier : A semiconductor detector utilizing a junction due to a surface inversion layer.

Sensitive volume of a detector : That part of the volume of a radiation detector in which radiation can be detected.

Sensitivity : In diagnostic tests, the number of test results registered as true positives in relation to the actual number of positives in the group tested. The minimum signal that can be detected satisfactorily. The lowest concentration of test substance measurable in an assay, or the minimal difference in test substance concentrations distinguishable by a given assay, usually a function of the steepness of the assay curve.

Separation energy : Of a particle, the energy required to remove that particle (e. g. , neutron, proton, alpha) from the nucleus.

Separation, radiochemical : See radiochemical separation .

Sequestering agent : Or sequestrates, are compounds capable of binding metal ions so that they no longer exhibit their normal reactions in the presence of precipitating agents. Most sequestrates owe their action to ?the formation, with the metal ion, of stable coordination compounds of chelates.

Serial dilution : The progressive dilution, by the same factor, of standard or sample in a row of tubes so that the first tube contains the highest concentration of test substance.

Series, radioactive : See decay chain .

Sf : Spontaneous fission.

Shadowing : The local reduction of the particle flux density due to the presence of a nearby absorber.

Shielding : Protecting against penetrating radiation by using an absorbing layer of matter. Also, such matter itself. *OR* A protective barrier, usually a dense material, that reduces the passage of radiation from radioactive materials to the surroundings by absorbing it.

Si : systems international

Si(li) detector : A lithium-drifted silicon detector. A diffused junction semiconductor detector in which silicon is the base solid state material.

Sievert : Unit of dose equivalent. (Not capitalized. Symbol =Sv.) 1 sievert = 1 gray X quality factor .

Sievert (sv) : A measure of dose (technically, dose equivalent) deposited in body tissue, averaged over the body. Such a dose would be caused by an exposure imparted by ionizing x-ray or gamma radiation undergoing an energy loss of 1 joule per kilogram of body tissue (l gray). One sievert is equivalent to 100 rem.

Sigmoid curve : S-shaped curve, approaching linearity in the middle and curved at either end.

Single channel pulse height analyzer : See pulse amplitude selector .

Single comparator method : In quantitative activation analysis, a method in which a known amount of activatable isotope is irradiated simultaneously and under identical conditions with an unknown amount of the isotope.

Snap : Space Nuclear Auxiliary Power; a radioactive source for power generation in satellites and space craft.

Snr : Signal-to-noise ratio.

Soft radiation : Radiation whose particles or photons have a low energy and consequently do not penetrate materials easily.

Solid state track detector : An insulating solid in which the paths of nuclear particles suffer radiation damage of sufficient permanency that they may be rendered visible through chemical etching processes.

Solute : The minor component of a solution which is regarded as having been dissolved by the solvent .

Solvent : A liquid (usually the major component of a solution) which is used to dissolve a solute or solutes.

Solvent extraction : Process of removing one or more constituents of a mixture from immiscible solvents by preferential solubility in a specific solvent.

Sorbent : Material which has a large capacity for absorbing moisture or other gas or fluid.

Sorption : It is sometimes difficult or impossible to discriminate experimentally between adsorption and absorption : in such cases it is convenient to use the non-committal term sorption. In ion exchange, sorption is the uptake of electrolytes or non-electrolytes by ion exchangers through mechanisms other than pure ion exchange.

Source : A radioactive material that produces radiation for experimental or industrial use.

Source efficiency : See self-absorption factor.

Source strength : The strength of a radioactive source, meaning the amount of radioactive material contained in it, usually expressed in curies or becquerels.

Source, extended : In contrast to a point source , below, a source of radiation whose size is not constrained to approximate a point.

Source, point : A source of radiation which is confined to a small point or spot.

Source, radioactive : Any quantity of radioactive material which is intended for use as a source of ionizing radiation .

Source, sealed : See sealed source .

Source, voluminous : See source, extended .

Spallation product : A product of a spallation reaction .

Spallation reaction : A nuclear reaction in which the projectile induces the ejection of neutrons, protons and/or light heavy ions from the struck target nucleus.

Special nuclear material : Fissionable and other material related to military use.

Speciation : A term that refers to the physicochemical forms of a radionuclide under a particular set of ambient conditions (pH, Eh, ligands present, etc.).

Specific activity : See activity, specific .

Specific burn-up : See burn-up, specific .

Specific energy : The energy imparted by ionizing radiation to matter.

Specific ionization : The number of ion pairs formed per unit distance along the track of an ionizing particle passing through matter.

Specificity : In radioimmunoassays , a measure of the discrimination between antigens by the antigen-antibody site of the antibody .

Spect : Single photon emission computerized tomography.

Spectral hardening : See hardening .

Spectrogram : A spectrum as recorded by a spectrometer .

Spectrograph : Graphical representation of a spectrum.

Spectrometer : A measuring assembly used for determining the spectrum of a radiation.

Spectrometer, alpha (,beta, gamma-ray, x-ray) : A measuring assembly incorporating a radiation detector and a pulse amplitude analyzer , used for determining the energy

		spectrum of alpha (beta, gamma, x) radiation.
Spectrometer, anti-compton gamma-ray	:	See gamma-ray spectrometer, anti-Compton.
Spectrometer, crystal diffraction	:	A spectrometer , in which diffraction by a crystal is used to obtain the energy spectra of electromagnetic radiation as well as of slow neutrons.
Spectrometer, scintillation	:	See scintillation spectrometer .
Spectrometer, x-ray	:	See spectrometer, alpha (beta, gamma-ray, x-ray).
Spectrum (,radiation)	:	The components of radiation arranged in order of their wavelengths, frequencies or quantum energies.
Spectrum analysis	:	The interpretation of the information present in an energy spectrum in terms of radiation energy and intensity.
Spectrum, measured	:	See spectrogram .
Spike	:	The isotopically distinct label (radioactive or stable) added to a sample undergoing analysis by any of several isotopic methods.
Spiking	:	The process of adding a spike to a sample undergoing analysis.
Spill	:	A contamination as a result of an accident during some process or procedure involving radioactivity.
Spin, nuclear	:	The intrinsic angular momentum of an elementary particle or system of particles such as a nucleus.
Spinthariscope	:	An instrument for viewing scintillations of charged particles on a luminescent screen, usually with the aid of a microscope.
Spontaneous fission	:	See fission, spontaneous .
Sra	:	See sub stoichiometric radiochemical analysis.

Srcs : See scattering recoil coincidence spectrometry.

Sse-ida : See sub- and super equivalence isotope dilution analysis.

Sstd : Solid state track detector.

Stable : Not radioactive.

Standard material : A reference material for which, for specified element concentrations, values are recommended by some official body. These values should be based on the consistent results obtained by using independent analytical techniques.

Standard model : Gauge theory encompassing the electroweak and strong interactions.

Standard reference material : Term exclusively used for a standard material issued by the National Bureau of Standards, USA.

Star : A nuclear event, recorded in photographic emulsion , in which the tracks of three or more charged particles radiate from a single point.

Stopping power : Of a substance, for charged particles of specified energy, the average energy loss in passing through a thin layer of that substance, divided by the thickness of that layer.

Straggling : A term used to express the fact that the ranges of charged particles of a given energy in matter are not constant but show a small spread about the mean range. This uncertainty in the range occurs because the loss of energy by charged particles is not a continuous process, but one in which energy is transferred in discrete amounts to the electrons in the matter through which the particle is passing.

Stripping : The process of back-extraction of a substance from an extract, usually into an aqueous phase.

Strong interaction : The interaction due to exchange of color. Also called strong force.

Strontium unit : A measure of the concentration of ^{90}Sr in food and in the body measured relative to calcium. One strontium unit is 1 millibecquerel ^{90}Sr per gram calcium.

Sub- and **superequivalence isotope dilution** : A technique of sub stoichiometric radiochemical analysis in which two series of solutions containing sample and spike are compared; one is isotopically diluted with increasing amounts of carrier and the other has multiples of sample and spike but no carrier. The amount of unknown element in the sample is obtained by graphic analysis following activity ratio determinations.

Substoichiometric radiochemical analysis : An isotope dilution technique combining A radioactive tracer and a reagent which is intentionally applied in amounts that are insufficient to react completely with the analyte.

Sub-trace analysis : Trace analysis with sample size 10^{-3}- 10^{-4} g and constituent content 100 ppm (0.01 %).

Sum peak : In a spectrum of radiation , the part of the response corresponding to the total absorption in a detecting material of the combined energies of two coincident radiations .

Sum peak method : A method of analysis in which the sum peak of two gamma rays is counted and in which the intensity ratios of the sum peak to the single peak can be used and changes in the ratios due to the chemical environments of the emitting nuclide can be observed.

Superequivalence isotope dilution : See sub- and super equivalence isotope dilution.

Superheavy element : A term often used with reference to elements of atomic number well beyond 110.

Supernatant : The clear solution associated with a settled, centrifuged, or filtered precipitate.

Surface barrier semiconductor detector : See semiconductor detector, surface barrier .

Survey meter : A battery-powered portable device containing a gas ionization chamber or other detection device for monitoring radiation levels.

Sv : Symbol for the unit, the sievert.

Sv : Sievert (SI unit of radioactive dose equivalent)

Symmetry : Invariance of equations of motion under changes in condition.

Synchrocyclotron : An ion accelerator, the chief components and configurations of which are similar to those of a cyclotron and in which the phase of the accelerating potential is synchronized with the frequency of the accelerated particles by frequency modulation to compensate for relativistic increases in mass at high speeds.

Synchrotron : An accelerator in which ions are accelerated around a fixed circular path by a radiofrequency potential and maintained in the path by a magnetic field that varies with time.

Synchrotron radiation : Electromagnetic radiation produced by any charged particle forced to travel in a nonlinear orbit such as a proton in a storage ring.

Synergism : A characteristic exhibited when two separate agents in combination produce an effect greater than the sum of that produced by each agent separately.

Synroc : The name given to a group of specially formulated zirconium-based ceramics that were originally developed for immobilizing high-level (radioactive) waste.

Szilard-chalmers effect : The rupture of the chemical bond between an atom and a molecule of which the atom is a part, as a result of a nuclear reaction of that atom.

Tac : Time-to-amplitude converter.

Tagged : See labeled.

Tandem accelerator : A two-staged accelerator, usually applied to accelerators of the Van de Graaff variety.

Target : The material subjected to bombardment by radiation, high-energy particles, or high-energy nuclei for the purpose of producing a nuclear reaction.

Targetry : The design of target composition and structure to optimize the production of a desired nuclide and minimize impurities made through nuclear reactions.

Tbp : Tributyl phosphate, a solvent used in extraction separations.

Tdmes : Time differential Mossbauer emission spectroscopy.

Tdpac : Time differential perturbed angular correlation measurement.

Thermal column : A large body of moderator , adjacent to or inside a reactor to provide thermal neutrons for experiments.

Thermal energy : Random kinetic energy possessed by objects in a material at finite temperature.

Thermal fission : See fission, thermal.

Thermal ionization mass spectrometry : A high-precision mass spectroscopic technique for the analysis of metal ions in which the sample is applied to a metal strip to which thermal energy is applied for ionization.

Thermochromatogram : The response of a detection system to a separation by thermo chromatography.

Thermochromatograph : A device for performing a separation by thermo chromatography.

Thermochromatography : A column chromatographic separation technique based on the variation with temperature of the affinity of different chemical species in the analyte towards an absorbent which lines the column.

Thermoluminescence : Luminescence produced in certain substances by heating.

Thermoluminescent detector : A type of crystal used to monitor radiation exposure by emitting light; often used in film badges or ring badges.

Thick target yield : The yield or reaction rate per incident beam particle for a thick target, that is a target in which the attenuation of the beam through the target along the beam direction cannot be neglected.

Thorium series : Natural decay chain commencing with ^{232}Th and ending with stable ^{208}Pb. The mass number of each member of the series is representable as 4n where n is an integer.

Threshold energy : See energy, threshold.

Time constant : 1) A convenient measure of the rate of a capacitor charging or discharging; namely, the time for the current to have reached a fraction, 1/e, of its initial value. 2) The time needed for a detector response to reach a fraction, 1/e, of its final value if the detector is suddenly exposed to a radiation flux.

Time-of-flight : A method of mass analysis based on measuring the time needed for an ion to traverse a known distance between source and detector.

Time-to-amplitude converter (tac) : An electronic instrument which converts the time interval between two timing (logic) signals into an output pulse with an amplitude that is proportional to the time interval.

Titer or **titre (in radioimmunoassays)** : A measure of the level of antibodies .

Titration, radiometric : See radiometric titration .

Tld : Thermo luminescent dosimeter

Tnt : trinitrotoluene

Tof : Time-of-flight .

Total absorption peak : See full energy peak .

Townsend avalanche : In a proportional counter, the gas multiplication cascade of secondary electrons.

Tracer : A small amount of radioactive isotope introduced into a system in order to follow the behavior of some component of that system *OR* Labeled member of a population used to measure certain properties of that population.

Tracer, generally labeled : A tracer in which the position of the label is not defined.

Tracer, intrinsic : An isotope, present naturally in a given sample, that may be used to trace a given element through chemical and physical processes.

Tracer, isotopic : A tracer which only differs in isotopic composition from the substance to be traced.

Tracer, nominally labeled : A tracer in which the label is present in a specified position.

Tracer, specifically labeled : A tracer in which the label is present in a specified position.

Tracer, stereospecifically labeled : A tracer in which the label is present in a stereo specific position.

Tracer, uniformly labeled : A tracer in which the label is uniformly distributed over its possible positions.

Track detector, nuclear : A detector which makes the paths of ionizing particles visible, either directly (e.g. cloud chamber) or after suitable treatment (photographic emulsion, polymers).

Track etching : For exposed solid state track detectors, the process of making the radiation damaged particle paths visible by chemically removing the material along the path.

Track, nuclear : Of an ionizing particle, its path as revealed by a track detector.

Transactinides : Elements beyond the actinide series, that is, elements with atomic numbers greater than 103.

Transactinoids : See transactinides .

Transformation, nuclear : The change of one nuclide into another.

Transient equilibrium : See equilibrium, transient.

Transition, isomeric : See isomeric transition.

Transition, nuclear : For a nucleus a change from one state into another; a nuclear transformation.

Transmutation : 1) Any change of one element into another. 2) Any change of one nuclide into another. ***OR*** The transformation of one element into another by a nuclear reaction.

Triga reactor : A nuclear reactor with uranium-zirconium-hydride fuel elements which usually operates at a steady state level but which can be made to generate very short pulses with neutron fluxes thousands of times greater than the steady state level. It is useful for pulsed reactor activation analysis.

Triple-monitor-method : In neutron activation analysis the use of a Au monitor in addition to the bi-isotopic monitors for flux parameter calibrations.

Tritiation : Hydrogenation using tritium-labeled hydrogen.

Tritium (breeding) ratio : The number of tritons produced per fusion event.

Triton : The nucleus of hydrogen with mass number three; the tritium nucleus.

Tta : Thenoyltrifluoracetone, a chelating agent commonly used in radiochemical separations.

U : Uranium

Ultraviolet radiation : Electromagnetic radiation having wavelengths between the visible part of the spectrum and x-rays.

Unsealed source : A radioactive source which is not encapsulated, in contrast to a sealed source .

Unstable : Radioactive.

Uranium series : The natural decay chain commencing with ^{238}U and ending with stable ^{206}Pb. The mass number of each member of the series is representable as 4n+2 where n is an integer.

Van de graaff accelerator : Device using a high voltage terminal to accelerate charged particles. OR An electrostatic generator in which a high potential is produced by the accumulation of electric charge conveyed to an insulated conductor by a continuously moving belt. The potential generated is used for accelerating charged particles.

Waste, high-level : (i) The highly radioactive liquid, containing mainly fission products, as well as some actinides , which is separated during chemical reprocessing of irradiated fuel (aqueous) waste from the first solvent extraction cycle and those waste streams combined with it. (ii) Spent reactor fuel, if it is declared a waste. (iii) Any other waste with a radioactivity level comparable to (i) or (ii).

Waste, low-level : Waste which, because of its low radionuclide content, does not require shielding during normal handling and transportation.

Weak interaction : The interaction responsible for weak decays of particles, mediated by the exchange of W± and Z0 gauge bosons.

Westcott cross-section : See cross section, effective thermal.

Wigner effect : Displacements of atoms from their positions in a crystal structure by radiation.

Wilzbach labeling : See labeling, Wilzbach.

X-radiation : Electromagnetic radiation usually produced in transitions of the inner electrons of atoms. The wavelength is between ultraviolet and gamma rays.

X-radiation, characteristic : See characteristic X-radiation.

X-ray : Electromagnetic radiation with wavelengths between ultraviolet and gamma rays. OR Electromagnetic radiation of wavelengths in the approximate range 10^{-11} to 10^{-9} m whose origin is in there arrangement of electronic states of atoms from an excited state to a lower energy excited state or to the ground state.

X-ray emission analysis, particle induced : A method of analysis based on the measurement of the energies and intensities of characteristic X-radiation emitted by a sample during irradiation with heavy charged particles.

X-ray emission analysis, radioisotope induced : A method of analysis based on the measurement of the energies and intensities of characteristic X-radiation emitted by a sample during irradiation with a radioactive source.

X-ray fluorescence : The emission of X-ray radiation by an atom as a result of the interaction of radiation with its orbital electrons.

X-ray fluorescence analysis : A method of analysis based on the ?measurement of the energies and intensities of characteristic X-radiation emitted by a sample during irradiation with X-rays of wavelength shorter than their ?characteristic wavelengths.

X-ray fluorescence analysis, energy-dispersive : A method of X-ray fluorescence analysis involving the measurement of the wavelength (energy) spectrum of the emitted radiation , e.g., by a semiconductor detector.

X-ray fluorescence analysis, wavelength-dispersive : A method of X-ray fluorescence analysis involving the measurement of the wavelength (energy) spectrum of the emitted radiation , e.g., by using a diffraction grating or crystal.

X-ray fluorescence analysis, wavelength-dispersive : A method of X-ray fluorescence analysis involving the measurement of the wavelength (energy) spectrum of the emitted radiation , e.g., by using a diffraction grating or crystal.

Xrf : X-ray fluorescence .

Yield, fluorescence : For a given excited state of a specified atom, the ratio of the number of excited atoms which emit a photon to the total number of excited atoms.

Yield, radiochemical : See radiochemical yield

: Planck's constant, h , divided by 2.

(ft) VALUE : The product of the half-life of a beta-decaying nuclide and a function (the Fermi function) that depends on the decay energy and transition type.

µ Z/a : micro (10^{-6}).

µCi : micro Curie.

Useful Quantities in Nuclear Science

Quantity	Symbol	Value
Atomic Mass Unit (U)	mass ^{12}C	$1.6605402 \times 10^{-27}$ kg
Electric Charge (Coulomb)	e	$1.60217733 \times 10^{-19}$ C
Speed Of Light (Vacuum)	c	299792458 m/s (exact)

Particle	Mass	Charge	Half-life
	(in u)	(in electron charge, e)	(in seconds)
proton, p	1.007276	+1	stable
neutron, n	1.008665	0	624
electron, e or e^-	0.000549	-1	stable
neutrino, ν	> 0	0	stable*
antiproton, $\bar{p}$	1.007276	-1	stable†
antineutron, $\bar{n}$	1.008665	0	624†
positron, e^+	0.000549	+1	stable
antineutrino, $\bar{\nu}$	> 0	0	stable*
muon, $\mu \pm$	0.10565	+1, -1	3.17×10^{-6}
pion, $\pi \pm$	0.14	+1, -1	3.76×10^{-8}
pion, π_o	0.14	0	6.63×10^{-17}
photon, y	0	0	stable
gluon, g	0	0	
up quark, u	approximately 0.005	+2/3	
down quark, d	approximately 0.01	-1/3	
strange quark, s	0.1 to 0.3	-1/3	
charm quark, c	1.0 to 1.6	+2/3	
bottom quark, b	4.1 to 4.5	-1/3	
top quark, t	180 ± 12	+2/3	
^{1}H atom	1.007825	0	stable
^{2}H nucleus	2.013553	+1	stable
^{2}H atom	2.014102	0	stable
^{3}H nucleus	3.015500	+1	3.88×10^{8}
^{3}H atom	3.016049	0	3.88×10^{8}
^{3}He nucleus	3.014932	+2	stable
^{3}He atom	3.016029	0	stable
^{4}He nucleus	4.001506	+2	stable
^{4}He atom	4.002603	0	stable
^{12}C atom	12.000000	0	stable
^{238}U atom	238.050785	0	1.41×10^{17}

† Antiparticles are assumed to have same half-life as particles
*Neutrinos (antineutrinos) can change into an other one of the same type.

From the Review of Particle Properties, R. M. Barnett *et al.*, Physical Review D54, 1 (1996).

International System of Units (SI)–SI prefixes

Value	Name	Symbol
10^{-24}	yocto	y
10^{-21}	zepto	z
10^{-18}	atto	a
10^{-15}	femto	f
10^{-12}	pico	p
10^{-9}	nano	n
10^{-6}	micro	μ
10^{-3}	milli	m
10^{-2}	centi	c
10^{-1}	deci	d
10	deca	da
10^{2}	hecto	h
10^{3}	kilo	k
10^{6}	mega	M
10^{9}	giga	G
10^{12}	tera	T
10^{15}	peta	P
10^{18}	exa	E
10^{21}	zetta	Z
10^{24}	yotta	Y

□□□

Appendix

Nobel Prizes in Nuclear Science

Many Nobel Prizes have been awarded for nuclear research and instrumentation. The field has spun off: particle physics, nuclear astrophysics, nuclear power reactors, nuclear medicine, and nuclear weapons. Understanding how the nucleus works and applying that knowledge to technology has been one of the most significant accomplishments of twentieth century scientific research. Each prize was awarded for physics unless otherwise noted.

Name(s)	Discovery	Year
Henri Becquerel, Pierre Curie, and Marie Curie	Discovered spontaneous radioactivity	1903
Ernest Rutherford	Work on the disintegration of the elements and chemistry of radioactive elements	1908 (chem)
Marie Curie	Discovery of radium and polonium	1911 (chem)
Frederick Soddy	Work on chemistry of radioactive substances including the origin and nature of radioactive isotopes	1921 (chem)
Francis Aston	Discovery of isotopes in many non-radioactive elements, also enunciated the whole-number rule of atomic masses	1922 (chem)
Charles Wilson	Development of the cloud chamber for detecting charged particles	1927
Harold Urey	Discovery of heavy hydrogen (deuterium)	1934 (chem)
Frederic Joliot and Irene Joliot-Curie	Synthesis of several new radioactive elements	1935 (chem)
James Chadwick	Discovery of the neutron	1935

Contd...

Carl David Anderson	Discovery of the positron	1936
Enrico Fermi	New radioactive elements produced by neutron irradiation	1938
Ernest Lawrence	Invented the cyclotron	1939
George De Hevesy	Use of isotopes as tracers in the study of chemical processes	1943 (chem)
Otto Hahn	Discovered fission of massive nuclei	1944 (chem)
Patrick Blackett	Improved cloud chamber and discoveries in nuclear physics and cosmic rays	1948
Hideki Yukawa	Predicted the existence of mesons as the basisof the nuclear force	1949
Cecil Powell	Developed the photographic method of studying nuclear processes	1950
Edwin McMillan and Glenn Seaborg	Discoveries in the chemistries of the transuranium elements	1951 (chem)
John Cockcroft and Ernest Walton	Transmutation of nuclei by accelerated particles	1951
Felix Bloch and Edward Purcell	Measured magnetic fields in atomic nuclei (NMR)	1952
Walther Bothe	Analysis of cosmic radiation using the coincidence method	1954
Willard Libby	For his method to use ^{14}C for age determination	1960 (chem)
Robert Hofstadter	Studied nuclear structure with electron scattering	1961
Rudolf Mössbauer	Discovery of recoilless resonance absorption of gamma rays in nuclei	1961
Eugene Wigner	Application of symmetry principles to the nucleus	1963
Maria Goeppert-Mayer and Hans Jensen	Developed the nuclear shell model	1963
Hans Bethe	Developed the theory of nuclear reactions in stars	1967
Aage Bohr, Ben Mottelson, and James Rainwater	Developed the theory of collective states in nuclei	1975
Rosalind Yalow	Study of insulin using radioactive tracers	1977 (biology)
William Fowler	Studies on the formation of nuclear reactions which produce chemical elements in astrophysical processes	1983

□□□

References

Baghel, B. S., Gupta, N. and Tiwari, R. (2005). Effect of different doses of gamma radiation on self life of mango fruits. Cv. Dashehari. *J. Nuclear Agric. Biol.* 34(3-4): 141-147.

Bhattacharyya, B., Gurg, A. K. and Gautam R. D.((2004). Labelling of Trichogramma chilonis Ishii and its factitious host, corcyra cephalonica stainton with phosphorus-32 and its influence on the biological attributes of the parasitoid. *J. Nuclear Agric. Biol.* 33(3-4): 198-205.

Deb, D. L. (1997). Application of isotopes and radiations to meet the new challenges in Indian Agriculture. *J. Nuclear Agric. Biol.* 26(1): 1-10.

Fried, M. (1964). "E", "L" and "A" values, Proc. 8th Int. Cong. Soil Sci., Bucharest, 29-39.

Fried, M. and Dean, L. A. (1952). A concept concerning the measurement of available soil nutrients. *Soil Sci.* 73: 263-271.

Girija, P. , Kumar, S. , Francis, V. and Abraham, C. T. (2006). Translocation studies in Loranthus(Dendophthoe falcta L. F.) using labeled 14C. *J. Nuclear Agric. Biol.* 35(1): 52-56.

Jagadeeswaran, R., Arulmozhiselvan, K. , Govindaswamy, M. and Murugappan, V. (2004). Studies on nitrogen use efficiency in turmeric using ^{15}N tagged urea. *J. Nuclear Agric. Biol.* 33(2): 67-76.

Jena, D. Misra, C. and Bandyopadhyay, K. K. (2003). Effect of integrated use of FYM and nitrogenous fertilizer on ammonia volatilizations loss and nitrogen use efficiency of wetland rice. *J. Nuclear Agric. Biol.* 32(2): 87-94.

Jha, P. Rattan, R. K. and Chandra, S. (2003). Influence of nitrification inhibitors, placement technique and irrigation sequences on utilization of labeled urea-N by wheat. *J. India Soc. Soil. Sci.* 51(4): 534-539.

Khatik, S. K., Deshmukh, K. K., Sharma, G. D. and Dwivedi, B. S. (2004). Effect or levels of P, FYM and PSB on mobilization of soil phosphorus using isotope technique in paddy. *J. Nuclear Agric. Biol.* 33(3-4): 163-169.

Kole, P. C. and Meher, S. K. (2005). Effect of gamma rays on some quantitative and qualitative characters in Zinnia in M_1 generation. *J. Nuclear Agric. Biol.* 34(2): 83-91.

Larsen, S and Sutton, C. D. (1963). The influence of soil volume on the absorption of soil P by plants and the determination of labile soil P. *Plant and soil.* 18: 77-84.

Larsen, S. (1952). The use of ^{32}P in studies on the uptake of P by plant. *Plant and soil.* 4: 1-10.

Larsen, S. (1967). Soil phosphorus. *Adv. Agron.* 19: 151-210.

Larsen, S. and Gunary, D. (1964). The determination of soil phosphates as influenced by the time of application of labelled phosphates. *Plant and soil.* 20: 135-142.

Mani, V. and Chandra, P. (2005). Effect of feeding γ-irradiated soybean on rumen fermentation pattern in goats. *J. Nuclear Agric. Biol.* 34(2): 115-119.

McAuliffe, C. D., Hall, S., Dean, L. A. and Hendricks, S. B. (1948). Exchange reaction between phosphates and soils. Hydrolytic surface of soil minerals. *Soil Sci. Soc. Ame. Proc.* 12: 119-123.

Paneth, F. and Vorwerk, W. (1922). Uber bine method zur Bestinmung der oberflache absorbierendez pulver. *Physic Chem.*101: 445-479.

Retheesh, P. K., Balachandram, P. V., Wahid, P. A. and Sachdev, M. S. (1997). Relative absorption of molecular urea and other forms of N by rice. *J. Nuclear Agric. Biol.* 26(2):105-110.

Russel, R. S., Russel, E. W. and Maraks, P. G. (1957). Factors affecting the ability of plants to absorb phosphates from soil. The relationship between labile phosphate and absorption. *J. Soil Sci.* 8: 248-267.

Semb, G., Olsen,R. and Steenberg, K. (1965). Comparison of different chemical and isotopic methods for estimating available P in solid. *Meld. Horg. L and brakagak,* 44: 21-23.

Sharma, K. N., Kaur, S. P. and Kumar, D. (2004). Effect of P fertilizers on their use efficiency in wheat in soils of different P status and their relationship. *J. Nuclear Agric. Biol.* 33(3-4): 135-145.

Spinks, J. W. T. and Barber, S. A. (1946). Study of fertilizer uptake using radioactive phosphorus. *Sci. Agri.* 27: 145-156.

Sud, K. C. (2005). Evaluation of crop residue on potassium kinetics in an acid soil and potassium use efficiency in potato-garlic sequence using tracer ^{86}Rb. *J. Nuclear Agric. Biol.* 34(3-4):128-140.

Thakur, R. Khatik, S. K. Deshmukh, K. K. and Sharma, G. D. (2005). Effect of FYM, SSP and Biofertlizer on utilization of phosphorus by rice in Typic Haplustert. *J. Nuclear Agric. Biol.* 34(3-4): 207-212.

❑❑❑

FSC
www.fsc.org
MIX
Papier | Fördert gute Waldnutzung
FSC® C083411

Zeitfracht Medien GmbH
Ferdinand-Jühlke-Straße 7
99095 Erfurt, Deutschland
produktsicherheit@kolibri360.de